FOREST REFORESTATION & RESTORATION: AN ANTHOLOGY

A society grows great when old men plant trees knowing they will never sit in their shade

FOREST REFORESTATION & RESTORATION: AN ANTHOLOGY

A.R.C.QUARTERLY FALL 1981

Journal of the Associated Reforestation Contractors of the Western United States

Inside: Workman's Compensation Update

By Dr. Bob Zybach

Published by NW Maps Co. Cottage Grove, Oregon USA

Est. 1994

Cover and title page photo of author by Bruce Fraser, Phoenix Communications, Inc.

CITE: Zybach, Bob 2025. *Forest Reforestation & Restoration: An Anthology.* NW Maps Co., Cottage Grove, Oregon USA: 136 pages. ISBN 978-1-7321276-9-2

FOREST REFORESTATION & RESTORATION: AN ANTHOLOGY

TABLE OF CONTENTS

TO THE READER:

This is the third in a projected series of four anthologies of my published articles, reports, maps, photos, interviews, and editorials from the past 45 years. The two previous collections focused on Elliott State Forest history, management, wildlife, and politics; and on the wildfire history and mitigation of western Oregon forestlands. This book focuses on Pacific Northwest forest reforestation and restoration histories; and the final collection will be about spotted owls, marbled murrelets, coho, their designated critical habitats, and the Endangered Species Act. They are written from the perspective of a forest scientist and amateur historian with strong opinions, based on extensive personal research and experience.

These collected writings contain significant scientific data in the forms of maps, tables, historical and documentary photographs, and the writings and spoken words of recognized experts -- but they have not been "peer reviewed" in the academic sense as that term is currently used; rather, the intended audience for these writings are my actual peers and reviewers: loggers, foresters, tree planters, roadbuilders, hunters, hikers, and fishermen. Rural businesses, residents, schoolkids, and grandparents. The people who have always lived, and worked, and played, in our western forests. True indigenous knowledge.

The intention of this series is to present the selected works in the format in which they were first published, including typos, ads, and even outdated or possibly incorrect information. The intended result is to retain original context and ambiance so much as possible.

Selections are mostly arranged chronologically, with the 1980 to 2010 timeframe beginning with an interview by Lono Waiwaiole for *Ruralite* magazine; and followed by several articles and photographs by me and by Bruce Fraser for *A.R.C. Quarterly*; and articles for Associated Oregon Loggers *The Log*; Oregon Small Woodlands Association *Northwest Woodlands*; and Association of Consulting Foresters *The Consultant*. The detailed interview by Dr. Brinda Sarathy regarding Oregon reforestation history and migrant labor was used for her PhD dissertation and for her book, *Pineros: Latino Labour and the Changing Face of Forestry in the Pacific Northwest*. The post-2010 timeframe is comprised of eight article/editorials written by me, and one by Dick Powell, for *Oregon Fish & Wildlife Journal* from 2012 to 2025, as formatted by editor Cristy Rein.

Bob Zybach
Creswell, Oregon
February 25, 2025

Ruralite®

POSTMASTER: Change of address & PO Form 3579 to be sent to Consumers Power, Inc., Bx 1108, Corvallis, OR 97330.

October 1980

Consumers Power

Oregon's oil!
Page 16

Oregon's oil!

Story and photos
by Lono Waiwaiole

Bob Zybach has struck oil in Eddyville.

Lots of oil—according to Bob, enough to let those foreign petroleum interests stew in their own oily juices.

Bob does business as Phoenix Reforestation. He has burned slash and planted trees all over the Northwest for most of the last 15 years, experience which has led him to several challenging conclusions. And the most challenging is that oil can grow on trees.

"That's not as crazy as it sounds," Bob says vigorously. "Think about it for a minute. This region is blessed with one of the most incredible resources on the planet. I wouldn't trade our forests for all the oil in Arabia—no way. They're going to run out of oil, and we don't have to ever run out of trees—not if we don't want to."

The challenge is in learning how to take full advantage of this incredible resource, a process that has been less than rapid. "How long have we been in these woods?" Bob asks. "A hundred years? One hundred and fifty? And so far, we've only scratched the surface of what can be done.

"Actually, we've gone backwards in some ways. A hundred years ago our forest-products industry wasn't based on petroleum like it is today, and we're not going to be able to afford a petroleum-based industry in the future.

"We are going to see some changes, that's for sure. There is no way that things can stay as they are. We know we have to become more self-sufficient. Our reliance on foreign oil warps our economy and our posture in the world, and self-sufficiency is our only long-term alternative. To me, that means learning to use what we already have."

And what we have in the Northwest is plainly evident from Bob's hillside just east of Eddyville—trees.

To the casual observer, Eddyville is just a wide spot on Highway 20 between Corvallis and Newport. To Bob, who is anything but casual, Eddyville is a good microcosm of forest management in the Northwest as it has been—and a good place to develop forest management as it must become.

"Today Eddyville is a post office, a school and a church. That's it. Phoenix Reforestation is the only business in town, and my crews work all over the Northwest so the local impact is very slight.

"But it wasn't always that way. There used to be a mill here. It stayed until all the prime timber was skimmed off, then the whole operation pulled out again. The old mill pond is still here, but that's about all.

"See that hillside across the road? That was logged off when the mill was here. It's replanted now, but that's very recent. Publishers Paper started buying up land a few years ago, and they have pretty much pointed the way to go around here as far as reclaiming the old logged-over land. I think what they are doing has opened a few eyes, and that's good.

"Publishers has distorted local economics, though. It's an old story. They buy land at more than the going rate, which arbitrarily raises property values. Then when a small-woodland owner dies, the heir can't afford the estate taxes—so he sells out to Publishers. It's a self-perpetuating cycle, and it's good for Publishers. But it's not good for small-woodland owners, and it may not be the most efficient way to manage the resource."

Bob rips through sentences like a power saw, his thoughts spraying in all directions. When he pauses to catch his breath—or to let his listener catch up—he tugs nervously on a well-developed Fu Manchu beard.

"That sort of bring us up to date," he begins again. "It's not a cycle peculiar to Eddyville, by any means. I've seen it all over this region. But let me show you what's going to happen next." He digs out a roll of tracing paper and spreads it out on the floor of the log house he must find the time to finish some day—some day soon, according to his wife.

On the paper is a sketch of the future of Eddyville as Bob Zybach sees it. It is a multi-faceted development which provides for local management of the local forest resources, all the way from seedling to sawdust.

"Right now the biggest challenge for small-woodland owners is competing with government and big corporations in an industry in which the product requires 40 years to mature. One of the biggest stumbling blocks has been finding good localized tree stock—the small-woodland owner just can't get it.

"So the first thing we need is a cooperative nursery here, so small-woodland owners can work together to develop their own seedling stock for reforestation. The cooperative aspect is the key, because working together we can utilize some of the same economies of scale that work for Publishers or Georgia Pacific.

"Together, we have enough clout to hold our own. Separately, let's face it—we're an endangered species.

"The second thing we need is a local mill that can take as small as four-inch tops. There is considerable timber here with commercial value right now if we can go down to four inches—and if we don't have to haul too far. Yaquina Veneer is ready to take care of this aspect.

"The third element is another key to the whole project, and that's a gasification plant that can use wastes from the mill and from the woods to heat everything—the greenhouse, the mill and some residences I hope to build on this shelf up here above the mill pond."

While he speaks, Bob is pointing everything out on the sketch. The mill, the nursery and the gasification plant will be along the railroad, not far from the site of the old mill.

"The railroad is another key. It gives us the ability to bring in solid waste to mix with the wood wastes—we think that will be an efficient blend for producing the gas. We hope to work with Larry Winiarski over in Philomath on this, because he's done a lot of work in the field and has the best process we've been able to find so far.

"My interest in all this is the nursery, plus building the houses if they are feasible. Did you know that solid, energy-efficient houses can be built for the price of a good mobile home? It can be done—we're doing it right now up near Siletz.

"Well, take a look at this house right here. This house will be here 300 years from now—no reason why it won't be. And it will be passive-solar when I get it finished. That deck will be a solar greenhouse, and we'll vent from it right into this room.

"The Siletz project is passive-solar, too. And up there we're using lumber milled right on the site from logs cut on the site. How much oil do you think that saves?

"That's what excites me about this project here in Eddyville, too. It makes sense economically. You can talk about energy alternatives all you want, but unless they pay for themselves they're never going to happen."

That attitude explains why Bob sheds no tears when OPEC raises oil prices. "Rising oil prices are the best thing that could happen to us. They force us to reevaluate what we're doing, and to look more realistically at the alternatives available. The reason we've disregarded the full potential of our forests in favor of imported oil is simple—oil has been cheaper.

"That won't be the case much longer, and it's a good thing—we've started to see some hidden costs in our oil habit, haven't we? There are lots of examples. If we didn't want Iranian oil, would Iran ever be in a position to hold Americans hostage?

"We have to turn to self-sufficiency, and what we're trying to do here locally is on the leading edge of that transition."

Bob's vision is more than lines on a roll of tracing paper. Most of the necessary land has been purchased, and the project is well on its way through the required maze of zoning ordinances and environmental permits.

However, not everyone shares his point of view. There is some local resistance to the changes he hopes will occur, and it may be stiff enough to eliminate Eddyville as the site. But Bob remains undismayed.

"This kind of project should happen somewhere. Everything is coming together—the motivation is there, the technology is there. All you need is a site with a forest base and maybe a railroad, and it really doesn't matter if it's Eddyville or Brazil. I just hope it happens here because this is where I live.

"See that Doug fir out there? What it all boils down to is kind of simple, really. That Doug fir is an oil well waiting to be tapped—and when you tap it, you can grow another one just like it.

"With a resource like that at our disposal, just how badly do we need Iran?"

Bob Zybach of Eddyville believes that oil can grow on trees, and he should know—he has been replanting the woodlands of the Northwest since his high-school days. "Our timber is one of the most incredible resources on the planet," Bob says, and he is working on a project in Eddyville that will make full use of that resource—all the way from seedling to sawdust.

Renewed Resources

The Reforestation of the Tillamook Burn (1948 - 1983)

By Bob Zybach

The Tillamook Burn is an important symbol to all reforestation contractors. Today the Tillamook Forest stands as a monument to the Oregonians who had the vision to finance the rejuvenation of a devastated wasteland and to the individuals and organizations who made that vision a reality.

The celebration of this fiftieth anniversary of the worst forest fire in modern history also recognizes the thirty-fifth anniversary of the November 2, 1948 decision by the voters of Oregon to pay for the reforestation of over 250,000 acres of prime timberland that came into state hands as a direct result of the 1933, 1939 and 1945 Tillamook fires. The vote on Article XI E, "Constitutional Amendment Authorizing Indebtedness for State Reforestation", was extremely close. Less than 2,00 votes of the total 420,500 cast decided the issue. Although it would be interesting to determine a rural/urban breakdown of the voting or to examine the votes cast from precincts directly affected by the burn, the remarkable thing is that the amendment was passed at all.

A 1944 Forest Service report stated: "The simplest approach is to regard the entire burn as beyond justifiable human effort . . . and let nature take it's course . . ." This "benign neglect" or "simplest approach" towards forest management is reflected today in management plans concerning Mt. St. Helens and in wilderness areas including commercial grade timberlands. In addition to the "justifiable human effort" hurdle, was the economic issue of whether reforestation was a justifiable investment at all. According to an article in the July, 1983 *Journal of Forestry*,

Many portions of the Tillamook Burn area contained magnificent trees such as these.

Notice the Willamette Valley in the Foreground of the 8 mile high mushroom cloud rising from the August, 1933 fire.

"until the war began, the cost of planting one-year-old Douglas Fir seedlings was twice the going price for well stocked 50-year-old stands". Although the argument has been made about the irony of the 1948 election being carried by early day "environmentalists", it seems more likely that a coalition of loggers, mill workers, bankers, politicians and others with a vested interest in Oregon's economy realized the need for a healthy crop of Douglas-fir to provide jobs for future generations. These people would more accurately be described as "futurists" than as "environmentalists".

In addition to the constitutional amendment passed in 1948, the same year saw the beginning of scientific studies concerning the soils of the Douglas-fir region. The first step taken in this direction was the funding of a teaching and research position in forest soils by the University of Washington. That September a Forest Soils Committee for the Douglas-fir region was established. The soil studies combined with the lessons learned from the mechanics and economics of regenerating the Tillamook Burn, led directly to reforestation contracting as we know it today.

A note of explanation is probably in order here. Most reforestation is accomplished by the natural re-seeding of harvested or devastated areas. In flat or sloping areas without desirable seed sources planting can be done better and cheaper mechanically. Douglas-fir, on the other hand, is not shade tolerant, usually grows on steep terrain and is far more valuable at an earlier age than other lubmer producing trees. These characteristics virtually guarantee that hand planting is a necessary step in obtaining the desired stocking levels for the establishment or regeneration of Douglas-fir. They also indicate that Douglas-fir is the most likely species to show a return for this type of investment. Part of the problem in assessing the economics of reforestation is whether the process is seen as a production cost, such as when fir are required by law to be planted following an old growth harvest, or as a strict long term investment cost, as when brush lands are purchased in order to be converted to Douglas-fir production for future industrial needs.

In 1949 the first contract for private work on the burn was put up for bid. Although Walt Milo of Hoodsport, Washington was the apparent low bidder at $14.40 per acre, Western Reforestation of Olympia obtained the job at $15.85 because Milo couldn't come up with the necessary bond. Bob Snow, a 1942 graduate of the University of Washington's School of Forestry and a partner with John Motishan in Western Reforestation, reports that this was hardly a major blow to Milo's fortunes; Walt Milo and his crews were later responsible for reforesting a substantial portion of the 1951 30,000 acre Forks Fire on the west side of the Olympics.

The first Tillamook specs were 6x8 spacing of 1-0 Douglas-fir that were so small that Snow says a planter could easily fit 800 or 900 in a small tree bag. With virtually no inspection and ground burned so clean that it "was like planting a garden", predominately green crews of young men from Forest Grove were able to average 1400 trees per man per day throughout the course of the contract. At approximately 900 trees per acre that would work out to about $24.00 per man day. The thing to remember is that wages were a lot closer to a dollar an hour at that time, social security and worker's comp were not real problems, and a new truck cost about as much as a small motorcyle does now.

In 1951 Bob separated from his partner in order to form Snow Reforestation out of his Portland base. Under that name his business completed a large share of the planting given to private contractors in the burn. Through the years his crews performed several large scale jobs for other landowners such as Crown Zellerbach and BLM in the burn, and in other areas such as the Siuslaw National Forest, the Olympics, and the Cascades.

In the early fifties other land managers began realizing the advantages of hand planting Douglas-fir as well as other conifer species, and several new businesses were formed or expanded their operations. Although most crews traveled from the first to extend their seasons, the successful businesses centered their activities around

Snags following fire.

a specific area. Snow in the Tillamook burn, Milo in the Olympics, Tom McCready and Owen Thomas in Medford, John Woods, Sr. out of Salem and L.A. Young and Wayne Young in Myrtle Creek.

This was a time of learning. As Tom McCready pointed out, several bench marks were established by the work to reforest the burn. In addition to the systems developed to make large scale plantations possible, several advances were made in critical areas such as seed extraction and seed storage. After the first plantations were established with seed gathered "from anywhere they could get it", and grown at the Nisqually I.F.A. nursery, it soon became apparent that seed source considerations and genetics would have to be given more thorough attention.

Statistics on the burn are everywhere. A list of firsts, mosts, and bests, concerning the devastation and reforestation of its 355,000 acrews would be literally endless. The Tillamook forest is a gauge. It is the result of a massive sotial enterprise that was successful; an experiment in intensive management without equal. Since the August night in 1933, when a 15 mile wide wall of flames pushed by 30 mile an hour east winds engulfed 220,000 acrews of standing timber in 20 hours, putting up a mushroom cloud 40 miles wide and 8 miles high, the burn has provided a mark to put comparable events and processes into perspective. Today we can easily acquire new knowledge from the burn. One hundred years from now our descendants will be able to do the same.

On July 18, 1973, 24 years to the day of the authorization of the reforestation bond, Governor Tom McCall officially dedicated what had been known for nearly 40 years as the Tillamook Burn, the "Tillamook Forest". On April 23rd, 10 years later, State Forester Mike Miller accepted a National Arbor Day award in Nebraska City, Nebraska, on behalf of the people of Oregon. The award recognizes the 50th anniversary of the 1933 burn, as well as those who participated in and supported its reforestation. With all of the attention, it is easy to forget that the Tillamook forest would be substantially different today if intense management had not taken place. It would be interesting to speculate what Tillamook County would be like today, if the Tillamook burn had been declared a "wilderness area", or what would have determined the futures of the Forks Fire or the Sundance Burn in northern Idaho, if Tillamook had not preceeded them. In any case, the $12,,00,000 spent by Oregonians between 1948 and 1973 had an effect, a large effect.

The title to this article, "Renewed Resources", is an obvious response to the well known bumper sticker declaring that "Trees are America's Renewable Resource." The reforestation of the Tillamook burn has resulted in changes to every single resource the forest has to offer. What are those changes and in what way do they benefit mankind?

The number one leisure time activity in the woods for most people is to simply drive through them in a car. A lot of people take part in this activity, and the vast majority seem to accept a trip along some back roads as a fulfilling, relaxing, enjoyable experience, worth sharing with friends and family. Although portions of the Tillamook Forest contained spectacular stands of old growth conifers before the fire, for the most part it was damp, in accessible country, with few game animals; only occassional breaks in the monotony of a continuous forest allowed scenic views of the Pacific Ocean or the Willamette Valley and the Cascades. There were few natural wonders worth visiting. When the smoke cleared people began to realize what a vast area was actually involved. Salvage logging operations opened it up. Seven billion of the ten billion feet given up for lost by the Department of Agriculture was salvaged. Green, fried or dried, the logs required roads to move them to the mills. Those roads were immediately used by hunters and fishermen, many of them loggers themselves. Foresters, reforestation workers, bikers, picnickers, hikers, students, ecologists, econimists, loners and lovers followed, and are still following many of the same roads today.

Clair Kunkel, Fish and Wildlife Biologist for the Tillamook Fish District since the late 1970's has some interesting observations concerning the effects of the burn and its reforestation on the migrant sport fish populations. Before the burn the old

Salvage operations proved boom to many businesses, prepared sites for later reforestation efforts.

Gales Creek, where burn started.

growth canopy transferred the energy from the sun to the fish bearing streams by shedding its growth in the form of needles and branches. Rotten snags, fallen trees and other forms of vegetation occasionally altered the streams solar diet and water flow. After the burn, solar energy was radiated directly into the water, altering its temperature, its aquatic life forms and its quantity. As snags began being harvested, as roads were built, and as fire-breaks were extended for miles, the physical condition of the streams was altered further. Besides the harvest roads being used by sports fishermen and meat fishermen, the commercial salmon fishing industry was growing rapidly. Yet the fish population was doing very well by the 1950's and was boosted to historically high levels in the 1960's with the aid of hatcheries. Clair believes that all of these physical effects upon the salmon bearing streams, while having a definite effect upon the health and population levels of the anadromous fish, are not as significant as the ocean conditions the fish encounter during the bulk of their lives. A glance at fish mortality statistics would seem to bear out this line of reasoning.

Using Coho salmon populations in Tillamook tributaries as an example, Clair recently explained to me the stream indexing system his department has used since 1950. At first, fish populations were measured at peak periods, but observations of high numbers of fish over short periods of time statistically out-weighing larger runs of lesser peaks but longer durations made observers adopt an average fish per mile index as more accurate than the peak index. He believes that seasonal and long term upswelling patterns of the ocean have extremely strong effects upon the survival of Coho smolts. In the case of stream bank devastation, such as occurred with the burn, native vegetation usually returns on coastal streams within a 3 to 8 year period to pre-devastation levels of influence. In the case of unseasonal ocean upswellings lasting for several years, Coho runs may take decades to recover. The present goal is an index of 29 Coho, or an average of 27 spawning Coho per stream mile during the duration of a run. Although peaks as high as 78 fish per mile were reached in the 1950's, the average was probably 25 ot 30. With hatchery aid, this figure hovered between 40 and 50 in the 1960's, a number that probably compares with any fish population preceeding the burn. Due to terrible ocean conditions this number dropped to 10 or 11 by 1977, and remains well below 20 to this day. The El Nino current doesn't seem to be helping.

Clair summarized his assessment of the effect of reforestation, logging, and fishing upon native fish runs, by pointing out the irony of people being able to file suit against the government in order ot "protect fish runs" at the same time the government, through regulation, does not allow scientists to measure the impact of the Marine Mammals Act upon fish populations. He also pointed out the Forest Practices Act of 1972 has proven to be an effective tool for wild fish managers and the net result may well be that current logging requirements are proving beneficial to salmon runs.

While Clair Kunkel believes that the burning and planting of the Tillamook watershed hasn't had an over-riding impact on fish populations, Doug Taylor, District Wildlife Biologist for the North Coast District for the past 7 years is keeping data on their impact on animal populations. While spotted owl observers, foresters, and lepidopterists all have an interest in the native life forms in a forest, the average individual is usually only interested in deer and elk populations, at least in the Douglas-fir region.

As with most forest fires eliminating an old growth cover, in a relatively short while, deer populations swelled in some areas to levels 100 to 400 times pre-fire levels. This is because Douglas-fir canopies are extremely shady so that most browse is eliminated once the trees fill in at an early age. Prior to the burn very few elk and deer lived in the area. It is very likely that other large mammal populations, such as bear, cougar, coyote, bobcat, and wolf were also very rare. Entries from Lewis and Clark's journals concerning the absence of large animal signs before reaching the elk herds of the Clatsop Plains bear this out. Even then they were forced to supplement the "poore elk meat" with dog meat, dried fish and a "fiew" roots traded from the local Indians in order to survive the winter. Once the canopy is removed, however, nutritious browse such as grasses, ferns, huckleberries, and tree seedlings almost immediately covers the forest floor. Deer populations explode. Elk populations seem to increase gradually. By 1953 the deer population in the burn had become so great that an inventory of seedlings established to that point showed 50% of them damaged by browsing. Special hunts were declared. The deer were so thick that a close relative of mine was able to fill 18 tags in the course of weekend hunts during the 1954 season. Doug thinks the special hunts were a case of "too little, too late" and had little, if any, impact on the deer population.

As the newly established forest began filling in in the mid-fifties, the deer population began declining. Although the severe winter of 1968-69 killed a large number of deer, their population rebounded swiftly in the next few years only to again begin the downward slide in numbers. At this time it seems that the deer population will continue to decline in the Tillamook forest until clear cutting operations begin in the next century or unless another disaster occurs that kills off substantial portions of the new trees.

Elk, on the other hand, have never occurred in the Tillamook drainages in any great numbers, at least not in the past several hundred years. Although elk populations increase at a much slower rate than deer populations, they are, for the most part, a hardier animal, better able to forage under more difficult circumstances than the deer. As a result, studies show the elk herds still increasing in size since the '68-'69 kill and quite likely to continue to increase through the remainder of this century. In concluding our conversation Taylor remarked that there are "probably more elk in the Tillamook forest now than there has ever been". Certainly far more than there ever were under an old growth canopy.

However, people did not pay 12 million dollars in an attempt to possibly better their fall hunt or to improve the esthetics of a charred wasteland. They did it mostly to preserve and create jobs. When the 6 year jinx struck again in 1951 the fire did not consume any new timber and was the last fire within the burn's boundaries. The fire control planning and work had proved to be a success. When Ken Risseuw fell and bucked a log this July it marked the beginning of what is conservatively estimated to be at least a one billion dollar return on the twelve million dollar reforestation investment.

Ken Risseuw operates Ken Risseuw Logging out of Sheridan. On February 18 of this year he bid $75/MBF, or about $35,000, on a two year contract calling for the commercial thinning of about 100 stems per acre on a 110 acre sale. The thinning is located 30 miles west of Forest Grove near the starting point of the original 1933 burn. It may well be the site of Bob Snow's first plantations from 1949-50.

By late September, Ken claims to be very satisfied with the progress of the sale with "about 50 loads decked" although trucking had not yet started. He says that the logistics of thinning a planted stand are "about the same" as logging a natural stand with the volume being the same or even slightly more than was figured when they bid the sale. Since the sale is based upon actual harvest volumes rather than "government estimates" he is not encountering the problems loggers are facing east of the Cascades on federal jobs where "optimistic" cruise figures, obtained at tax payer's expense, are causing severe hardships for many of the smaller contract logging firms. Despite the recent downturn in lumber prices Ken seems satisfied with this sale and claims that he would "bid about the same with the same conditions" if the opportunity were repeated.

The reforestation of the Tillamook burn has created many jobs where there were few, has provided much information where there was little and opened up vast tracts of real estate for recreational opportunities where there were none. Was it worth it? For Oregonians, loggers, hunters, silviculturists, students, and especially, for reforestation workers, the answer is an obvious and emphatic, "yes"! For environmentalists, preservationists and self-anointed ecologists and activists it certainly doesn't appear to be the "ecological disaster of mono-culture" fantasy that many of them seem hell-bent on promoting to an uneducated public. In fact, it rather appears to be the very opposite of what many of these "experts" seem intent upon promoting. The Tillamook Forest makes an obvious answer to these doom-sayers and makes one wonder as to their true motivations.

The Tillamook Forest, then, is many things to many people. Depending upon your perspective it can be viewed as a monument, a symbol, a gauge, an experiment, or an answer. But most of all it is a statement. It says, and will continue to say, a lot about man's ability to overcome incredible adversity and turn a disaster into an asset, rags into riches.

Fifty years from now people should be taking great pride in what has been accomplished with the Tillamook Burn. This year's rightful celebration of the reforestation of the Tillamook Forest should not lose sight of the fact that our work, far from being finished, has really just begun.

The author planted trees in the Tillamook Burn with this Grant H.S. (Portland) Biology class in 1963.

SAFETY AND PROFIT

by BOB ZYBACH

Unfortunately, one of the fastest ways to shift a worker's (or a contractor's) mind into neutral is to mention the words "safety" or "training" as if you were going to pursue the subject. I say "unfortunate" because I think that it can easily be demonstrated that in a service industry, such as ours, there is a direct link between safety and training, training and efficiency, and efficiency and profit. In other words, dollars invested by the contractor into safety and training programs should yield a profit. As with any investment, a great deal is dependent upon how those dollars are invested and of course upon a degree of luck.

Although the trend is varying, it has almost been traditional to hire reforestation workers through a phone call in answer to an ad or by a request sent to the local employment office. Thousands of workers will be hired this year merely through word of mouth or by showing up when the crummy arrives in the morning. No application. No physical. No training. This is O.T.J. Training in one of it's purer forms. Most of these workers will be transported to work sites in vehicles in direct violation of chapter 47 division 57 of the Oregon Occupational Safety and Health Code. At the job site many of these men will be given a chain saw for the first time in their lives and with the benefit of a little instruction (or none at all) turned loose thinning or slashing trees. A few of these men will die or be seriously injured.

In the past there has been little trouble with this system as the workers - many of them foreign and in the country illegally - have been reluctant to file civil suits for damages due to contractors negligence or to demand compensation insurance for job-related injuries. Those times are in the past. With the exception of a few contractors willing to exploit illegal aliens, several co-ops and some partnerships, most contractors are beginning to carry legally required amounts of compensation insurance. This legitimacy carries with it at least one burden: costly malingering, faked accidents, faked injuries and physical and mental injuries inherited from other employers.

• Which brings us to our first example, a Case History. Too bad it was mine. I hired a worker last fall that was highly recommended by his neighbor, one of my better planters. No questions, no writing (except a W-4), just Normal Hiring Practices. While the ground was scarified and the seedlings small he was everything his friend said; hard working, competent and reliable. When we took a two week Christmas planting break I put him on a cutting crew. After a couple of hours he said his hemorrhoids would keep him from doing any further cutting. This and a dubious case of malaria are the only excuses I've heard only one time. I should have been suspicious, but he'd been such a good worker I looked upon it as a humorous incident and little else. After a week of letting him set chokers (at a relaxed pace) on another job site we were ready to get back to planting, only this time we were planting the largest plug-1 seedlings we have ever handled. After a few hours work I saw him on the unit and asked him how it felt to be planting again. He said his back hurt. He's still on comp. Even though he has cost my company $11,000 this year, he has hired a lawyer to appeal for more. He had hurt his back in California a few years before for another employer. It had taken him nearly six months of "fighting" to secure his claim there, but this was Oregon. Here he was an Instant Winner. Other workers could not help but be impressed at the length of the paid vacation being enjoyed by this apparently healthy man.

A well-worded application form would have probably screened out this man as a potential health (safety) problem. And our company would have been returned several thousand dollars under our existing insurance plan if this "re-injury" had not occurred. At this time Chuck Burkey, Loss Control Representative for E.B.I., is working with Dean Pihlstrom, Bruce Fraser and I to develop a legal application form that our firms can use to help screen out workers with a greater than average potential to become injured on the job.

• My second example is more hypothetical and involves crew A and crew B, two pre-commercial thinning crews. Crew A features a foremam that periodically makes the rounds of the crew exhorting his men to work harder and to not mess with their saws: he will do that. If they dull their chain on a cable shortly after he leaves, they are still expected to keep working until he returns, at which time they are conditioned to really hump until he shuts them off. Then they ask him to sharpen their chain. As any experienced cutter knows, a dull chain will cause arm and back fatigue, will be more apt to "kick-back", more apt to bind or cause "barber-chairing," in fact, be more apt to do just about anything than cut fast. And yet the man will not even be given a file and will be told to do nothing on his saw but put the chain back on if it comes off. Crews like this really do exist. Obviously, this type of situation is more hazardous and less productive than it would be if the chains were sharpened immediately upon being dulled.

Crew B also has a foreman that periodically visits each of his cutters. Not only is he making sure that each of his workers is alive and steadily working, but he is also checking for safety gear such as gloves, good boots, hard hat and chaps, correct positioning in relation to the other cutters and the operating condition of the machinery. If he observes a man

Too late to back out - Bruce Nelson and Ernie Oleman.

faster, thereby adding to the potential to increase production. Each of these factors directly relates to a contractor's profit margin.

All other things being equal, it would seem that even the slightest amount of training (How To Sharpen Your Chain) would give an obvious bidding edge to crew B over crew A.

• To summarize these two points (and simultaneously attempt to avoid the nether void of "mods", "retro plans," and "deviated premiums") I have attempted to show how even the most modest of investments in the area of safety and training can realize good returns. In the first example, if I had invested a few hundred (or even a few thousand) dollars into a workable application form I could have screened out a potential injury fairly easily and earned several thousand dollars return on my present insurance plan. Or I could have placed the applicant in a position that would not have placed a strain on his back (if I had one). In the second example I have tried to show how even the most rudimentary instruction can almost immediately lead to a more productive (and profitable) man-day's work. Also safer.

I believe that a safer working environment, or even an attempt at a safer working environment, is good for crew morale and that morale is directly tied to productivity and work quality, often two of the most obvious characteristics of profitable operations.

cutting with a dull chain he immediately shuts him off and, if necessary, shows him how to sharpen it. After a few weeks even the newest men are operating their equipment near peak capacity. This is bound to lower the injury rate over the long haul, dramatically lower saw maintenance costs due to vibration caused by dull teeth (burned bars, clutch bearings, sprockets, etc.) and keep the saws cutting

PUBLICATIONS

There are a number of publications available to assist the reforestation contractor keep abreast of current market conditions, tax & law changes, job opportunities, and scientific studies. Here is where you can find a few.

National Federation of Independent Business Research and Education Foundation "Mandate"
150 West 20th Avenue
San Mateo, CA 94403

Forest Log
State of Oregon Dept. of Forestry
2600 State Street
97301

Timber Service and Marketing Report
P.O. Box 10162
Eugene, Oregon 97440

Forest Contractor's Weekly
1021 Forrester N.W.
Albuquerque, New Mexico 87102

Tree Planter Notes
U.S. Department of Agriculture
Pacific Northwest Forest and
Range Experiment Station
809 N.E. Sixth Avenue
Portland, Oregon 97232

Also, free copies are available from our Newport office of Oregon Dept. of Forestry's new brochure: "The New Reforestation Tax Incentives."

Douglas Fir Plug-1 Roots - G.P. stock

Phil Hahn's styro-block method of plug planting. The planter is John Avila.

Styro-blocks on the unit. The planter is Rocky Polries.

Reforestation: A Look Into The Future Of Contract Logging

A reforestation contractor takes a look at reforestation techniques and looks to the future with an eye to maximum yield through sound reforestation management practices.

by Bob Zybach

Bob Zybach is a reforestation contractor and a tree farm manager with 17 years experience in establishing and maintaining tree farms.

Although vast tracts of static second growth and old growth timber still exist in Oregon, most of it is located on publicly-owned lands that are either being "preserved" or else managed according to systems and formulas that guarantee a leisurely rate of harvest. Despite recent efforts to deviate from these "sustained yield" or "non-declining even flow" systems, it is possible that these lands will never be harvested in the economically viable manner of privately-owned timberlands. As a result, the contract-logger seeking job stability and financial security, rather than the "booms" provided by timber speculation, must look toward private lands and the crops they contain in order to speculate accurately as to the future of his business. How did this situation occur?

O & C Lands

In 1905, the National Forests adopted the sustained yield system. This was before the exhaustion of the timber supply east of the Rockies and before the opening of the Panama Canal in 1915. These two circumstances led to the first real demand for western timber, mostly for eastern construction products. The federally-owned commercial timberlands are located entirely in the west, with the most productive holdings located in Oregon.

In 1916, Congress passed the Chamberlin-Ferris Act which revested title of the remaining 2,900,000 acres granted to the O & C Railway in 1866 to the United States. This act classified the O & C lands into three categories. First, water-power sites were to be designated for disposal in the manner of similar government holdings.

Second, all 40-acre subdivisions containing a minimum average of 7½ MBF per acre were to be classified as timberlands and third, everything else was to be classified "agricultural." Through competitive bidding the act required that "...timber shall be sold as rapidly as reasonable prices can be secured therefore in a normal market." Following harvest, those lands would then be re-classified as agricultural. All lands classified as agricultural were opened to entry under the provisions of the homestead laws, except for a charge of $2.50 an acre.

On the timber holdings, this was to be collected from the log sales with the land being made available free to the homesteaders following harvest. These homesteaded lands were not allowed to be commuted. The $2.50 was divided between the counties, the state school fund, the reclamation fund and the general funds of the United States. Be-

The author in a stand of trees that were planted 30 years ago, left unmanaged until author thinned stand 3 years ago. Thinned, the trees are growing 2-to-4 times faster than BLM estimated cut projections.

tween 1916 and 1937, 375,000 acres passed back into private hands in this manner.

Congress Changes Act

In 1937, Congress passed a new act which placed the remaining O & C lands into the hands of federal land managers and removed them from settlement. The act stated explicitly that these lands were to be managed according to the principles of sustained yield. This was partially defined as "...contributing to the economic stability of local communities and industries." Allowable cut was determined to be 500,000 MBF per year until technical studies determined that the cut could be increased. By 1962, the cut figure stood at 1,127,000 MBF, although the average sales for that 25 year period consistantly averaged about 10% less than that provided for by law.

Since sustained yield charts were based directly upon natural growth patterns of timber rather than managed growth potentials and since such a large portion of Oregon (over ½) is owned by the federal government, it is easy to see how additional pressures were placed upon the private lands to increase harvest levels. When coupled with the subsidized housing and greatly increased bureaucracy following World War II, it often seemed as if the government was placing unfair demands upon the tax-producing private sector to provide the very crops it was protecting for itself. No one was complaining, of course, for these were boom times for the timber industries and it still seemed as if the timber was virtually limitless and impossible to harvest too fast.

Now, most of the private lands have been cut over and most projections seem to indicate that it will be about two decades before harvest schedules return to anywhere near their past levels. This abrupt, artificial halt in timber supply has caused most land managers to look closely at their regeneration plans and to realize that the primary value of their holdings was not in the timber that naturally existed upon them, but in the soils and weather patterns that produced the timber. As a result, virtually every major private timber producer has adopted some form of an intensive management program. Many of these programs have been simulated, at a great hidden expense to taxpayers, by various government agencies.

Cornerstones

The cornerstones of intensive management are logging and reforestation. If the land contains predominately merchantible trees, the first step is to determine property boundaries and cruise the holdings. Based upon these findings and also such things as market conditions; the Forest Practices Act of 1972, esthetics, other potential uses and future market conditions, harvest and regeneration plans are made. The land is then unitized and, if needed, road rights are secured to the individual units.

When the time comes to log a given unit, the next step is usually to design a permanent harvest road system within its boundaries. This system must consider permanent harvest needs as best as possible, as well as taking into account future reforestation plans. Using all of the above considerations, in addition to any immediate concerns such as available equipment and market shifts, roadbuilding and then logging are started. Upon their conclusion, the reforestation process generally begins.

If the land contains predominantly unmerchantible or sub-merchantible trees or brush, or is in need of some form of timber stand improvement (TSI) prior to any harvest considerations, then it is usually ready for some form of reforestation procedure. Again, the first step is to determine the property boundaries.

Although a cruise is generally unnecessary during this stage of management, some form of inventory or stand examination is usually made if it is determined to manage the existing trees to a commercial thin or clear cut stage. With the singular exception of TSI programs in which some merchantible material is harvested to partially off-set overhead or accrues during the enhancement procedures, road building or maintenance is rarely associated with reforestation practices. Using many of the same considerations listed above, the land is then unitized and necessary road rights obtained. At this point the land manager is generally ready to begin the implementation of his reforestation plan.

The next consideration is to determine the quality and type(s) of tree the manager wishes to establish on the unit. In the case of managing flat or gently sloping east side pine tree units, this is often done by the natural seeding of the remaining "leave" trees from a commercial thin or shelterwood approach to regeneration. In the case of clear cuts, it is usually necessary to obtain seed or seedlings through any one the several channels.

Seed Banks

Cooperatives such as the Industrial Forest Association exist that develop seed banks and seedlings for its members. Private concerns such as Simpson Timber near Crawfordsville and Tyee Nursery near Elkton exist that have an excellent record in dealing with land owners interested in developing and maintaining high grade seed banks and seedling supplies. In competition with these firms are federal nurseries, such as the one near Wind River and the D.L. Phipps Nursery, operated by the state of Oregon.

Since conifers only produce cones in sufficient quantities to justify harvesting sporadically (about 7 to 12 years for Douglas Fir), it is usually necessary for land owners to build up their reserves at those times. Cones are often purchased from local pickers or through hiring reforestation contractors during the "good cone years" and the seed is then processed, inventoried and stored for future years. This forms the basic seed bank and the seed, if stored properly, is good for at least 10 or 15 years. "Super" trees are developed by selecting cones

Proper use of a Hoe-dad.

from specific parent trees or by grafting scions from those parents onto root stock located in convenient sites. Through such methods as artificial pollination, establishing progeny test sites or even cloning, these trees can be improved genetically. Physical methods

(Continued on Page 28.)

(Continued from Page 27.)

of improvement include root pruning, fertilizing, transplanting, containerized growing and chemical treatments to reduce animal damage.

Site Preparation

As the seedlings are being developed for a specific site, the site itself is usually prepared in some manner so as to reduce competition from vegetation and animals, reverse soil compaction or to otherwise encourage the establishment of the seedlings. Good site preparation reduces future maintenance costs as well as immediate planting labor costs. Although site prep can involve such procedures as boomer (mountain beaver) trapping, hand brush piling or plowing, it is usually accomplished by the aerial application of herbicides, scarification or broadcast burning. Flat areas are generally scarified into piles or windrows that are often burned as flat terrain does not carry a flame well. Although steep slopes have been scarified through various means, this practice has been discontinued due to problems with erosion.

Broadcast burning is accomplished by fire-trailing the boundaries of a unit, slashing unmerchantible materials prior to, or following, a logging operation to eliminate unwanted vegetation and to provide fuel, spraying with a dessicant to reduce moisture, and then igniting by hand with drip torches or with jelly gas by helicopter. This is the most natural form of site prep and probably the best for long term management needs. It is also the most dangerous and potentially the most costly.

Planting

After the site is prepared it is then planted. This is done as quickly as possible to maximize the advantages provided by the preparation. The stock is almost always one-to-three years of age and is planted between September and June while it is dormant. Most planting is now done according to the micro-site method. The desired density, usually between 250 and 500 trees per acre, is determined by the land manager. The planter maintains an average grid of 9 to 14 feet, depending upon the requested density. Variations from this average are made on almost every seedling, depending upon such variables as soil density, slope, vegetation, required shading, animal signs, etc.

Once the planter has determined his exact planting spot, he then prepars it for the plant. This is usually accomplished through some form of "scalping," or removing undesired duff, slash, roots, snow or charcoal from the immediate area. The tree is planted in the scalp as centrally as possible.

There are four basic methods of planting seedlings by hand. Bare-root seedlings are usually planted by means of a hoe-dag, a shovel or a chain saw motor-powered auger. These trees are two or three years old and six inches to three feet in height. Plugs are planted at one year of age with a dibble or hoe-dag. Good planters will plant 500 to 2000 trees a day, depending upon contract specifications, the seedlings used, and the terrain and weather. The planting is continually monitored, as this is definately the most critical phase of any intensive management program.

SMR

Following planting, a wide variety of procedures insuring the establishment of the plantation can take place. Due to the potential diversity of these processes, my firm has taken to categorizing them all under the general heading of seedling maintenance and replacement, or SMR.

(Continued on Page 37.)

Reforestation

(Continued from Page 28.)

Included are bud-capping to discourage deer browse, staking and tubing to discourage boomer kill, aerial applications of herbicides, replanting in units with excessive mortality, hack and squirt, trapping, mechanical release, etc. Usually the land manager determines which process is to be done based upon stand exams, past history or the results of a continuous forest inventory (CFI).

After the seedlings are established, a term used when the leaders of the trees are above browse height and above competing vegetation (about five feet) and the basal diameter is large enough to discourage most damage by rodents, (about three-quarters of an inch), very little, if anything, is done for several years.

Once the canopy has closed in, however, a situation in which lateral growth from the trees has filled in most available space between stems, further treatment becomes necessary. Two things can occur at this point. In some areas the shade created by this condition becomes intense enough to kill most ground vegetation. Animals, such as the boomer, may then take to eating the bark on the saplings. If the tree is not killed by girdling, it at least is stunted by the damage and the open wound can be an entryway for any of several diseases. If this were the only problem associated with a closed canopy, the situation could be easily solved with a trapping program.

The basic problem, though, is that the trees begin competing directly with one another for all available light and moisture. If this situation is left untended it results in the addition of several years or even decades to the rotation cycle.

Thinning

Statistics seem to show that 200 trees will produce about the same fiber per acre over a period of time as 400 or 600 trees will. If so, it only stands to reason that the lesser number of trees will be cheaper to harvest, cheaper to move and worth more at the mill than the pecker poles, although there are a few exceptions to this rule. As a result, most land managers consider pre-commercial thinning (PCT) the most critical and most valuable investment, following planting.

Cross sections of tree stems will show that trees begin growing slower at the point that their canopy closes. By thinning to a desired density at that time, individual rapid growth continues. By waiting much beyond this point the trees become dependent upon one another for support against the weather, as top growth is not slowed appreciably by the closure. This is what is known as "dog hair." If the trees are thinned too late, they often become subject to wind and snow damage until they can recover their previous growth patterns and extend their roots into the opened areas. This often takes several years and is a critical phase for trees that are thinned late.

Most PCT programs were not implemented until the early seventies and foresters were often too conservative as to their desired specifications. The result has been that thousands of acres have needed to be re-thinned (an expensive process). Now trees are usually thinned between the ages of 9-to-15 years at a density of around 200 to 250 stems per acre. Besides reducing potential rodent damage and greatly increasing individual growth patterns, PCT has the distinct added advantage of removing deformed and diseased trees, undesired species and competitive brush, as well as favoring healthy, large and well-formed trees of desired species over their less fortunate neighbors. The humus provided by this process is probably also desirable, although it can form a potential fire hazard for a year or two. This is exactly the same approach taken by a vegetable gardener when weeding and thinning his carrots.

Staking and tubing is the most widely used method used for protecting young trees. It is costly, requires follow-up maintenance and the boomers still climb up and eat away the top of the tree.

Bud-capping is designed to reduce animal browse damage. This method is not 100% effective as photo indicates.

Additional TSI

Between PCT and a commercial thin or clear cut, situations can arise or specifications be desired in which further reforestation procedures are needed. These might include fertilizing a low-site stand, treating a mistle-toe infestation, removing bug or wind kill, etc. These procedures are known collectively as TSI (referred to earlier). Usually, though, TSI contracts are performed in marginally merchantible or sub-merchantible stands that have gone previously unmanaged.

Management roads might be constructed to remove an over-story of trees, suppressing the bulk of the stand, a multi-stage thinning program may be implemented to compensate for the lack of PCT as a way to increase stand quality without exposing the trees to sun scald, wind or snow damage excessively, pulp trees may be removed from a lodgepole stand in an attempt to make the leave trees vigorous enough to withstand an approaching beetle infestation, or brush may be piled or broadcast-burned to reduce a potential fire hazard. As a general rule of thumb, if most of the material removed is merchantible as logs, then the process is probably a logging procedure. If not, it is probably a reforestation procedure.

Obviously, the procedures out-lined above, if performed throughout the life of a stand (40-to-100 years), would be expensive. With the price of money today, it is easy to see how foresters are becoming more and more conscious of the economics of tree farm management. Investments totaling hundred of dollars per acre that are compounded for decades, must be made in as judicious a manner as possible. However, once a site has been prepared and a permanent

(Continued on Page 39.)

(Continued from Page 37.)

harvest road system installed, all further rotations become far less costly to manage.

Another factor is the crop itself. No more widow makers, dog hair, snags, or walls of salal. Mainly, just high grade, well-spaced, uniform trees with good access between them. The perceptive contract logger can easily see that as time goes on, he will be building less roads, paying less insurance, using smaller equipment (less payments) and having to be more efficient to compensate for these factors. This translates to greater investments in job training and accounting.

Brush and slash clean-up on the author's land. House in left background was built of beetle-killed pine entirely.

Where the ground is flat or gently sloping, he can see that the logger is becoming far more mechanized than he was even a decade ago. In these instances, additional investments into modern harvesting equipment is more than compensated by dramatically reduced payroll costs. Since the bulk of Oregon's log dollars are located on steep ground, however, it it unlikely that we will ever become as mechanized as the southern logging industry. Fallers and choker setters are apt to be with us for quite a while.

Forest's Future

We have looked at what the private forest of the future is going to be like and how the logger is likely to be affected, but what effect does reforestation have on the logger now? Besides the obvious changes in harvesting procedures implemented over the past several years, additional job opportunities are becoming available.

Many loggers use their large Cats to scarify units during the summer. Small Cats are used to fire-trail and brush pile in the summer and can be used to plant certain areas during the winter and spring. Snag falling is being performed more often than in the past and many cutters are slashing during seasonal down-time or in conjunction with the falling.

As managers see the needs of an improved harvest road system, additional roads are being built and existing roads are being re-opened and maintained more than ever. Loggers with access to helicopters are yarding slash to portable chippers, spraying herbicides, fertilizing, burning and transporting seedlings to areas with poor access.

During depressed times, such as we are currently experiencing, this potential for diversity is proving to be a great benefit to many contractors. As harvest levels increase, these procedures will guarantee a safer, more productive work environment in the decades to follow.

The Years Ahead

The next twenty years are projected as difficult ones for the Pacific Northwest timber industry. After that, there should be a resurgence of interest in forest products from this region, much as is being experienced in the south today. Reasonably projected demands for timber in the next century often make it appear as if the primary variable for the future of the logging industry is the stability of civilization itself.

Assuming present population trends, the present (although decreasing) per capita demands for wood fiber and the rapid development of new wood products by the forest industries, it seems probable that there will soon be a demand for all of the wood Oregon can produce. Modern reforestation practices are insuring that there will be trees when they are needed. And trees are the best job insurance there is for logging. Working together, we can obtain the maximum yield, now and forever, from Oregon's forests.

Facts and Opinions

WHAT IS IN A NAME?

Under the State of Oregon statutes any person or company who, for compensation, supplies workers for the planting of trees must be licensed to act as a tree planting contractor. When this law was passed in 1975, the Oregon legislature, for unknown reasons, combined tree planting contractors and farm labor contractors into the same license.

In 1981, the President of the ARC, Steve Winston, engaged Wendy Roberts, the State Labor Commissioner in conversation and tried to explain some of the problems that reforestation contractors encounter by being included with a different type of business; the farm labor contractors. After listening to the legitimate concerns as presented by Mr. Winston, the labor Commissioner quoted Shakespeare in summarizing her answer, "What is in a name?"

Later, while campaigning for re-election in 1981, one of the points the Labor Commissioner stressed in her campaign was that she was proud of changing the name of the "Oregon Bureau of Labor" to the "Oregon Bureau of Labor and Industries." She stated that the name change more accurately fit the responsibilities of the State Agency and therefore represented a valid effort on her part ot obtain that change.

It seems ironic that the administrator was sensitive about the name of the agency she heads, and yet appears to be so callous toward the problems of the industries she is elected to regulate.

Note: The latest effort on the part of the ARC to separate its licensing requirement from farm labor contractors was contained in a statement from a committee comprised of Brad Marks, Bob Zybach and Jim Stauffer. The statement commented on SB 525, the reforestation contractors licensing bill:

> Reforestation contractors do not employ migrant laborers in order to harvest seasonal crops, nor do they even work on agricultural "farms", therefore it is unreasonable to expect them to obtain a "farm labor contractor's license". Due to overwhelming documented evidence, however, it is obvious that there have been massive abuses of our present tax and insurance laws, among others, that point towards the need for a "reforestation contractor's license" in this state.

TERMINATION ON FEDERAL CONTRACT JUSTIFIED WHERE CONTRACTOR FAILS TO OBTAIN STATE TREE PLANTING LICENSE

On a pre-award survey for a Bureau of Land Management tree planting contract in a district in Oregon the Contracting Officer, among other requests, wanted to know if the contractor had secured the required State License for tree planting contractors. The contractor verbally verified that he did not have the license but was in the process of making application.

The contract was subsequently awarded to the Contractor which was a partnership of seven members. After award and the starting of the project by the contractor, the C.O. sent a telegraphic 10-day cure notice to the contractor. The partnership responded to the cure notice that they were unable to post a $5,000 labor bond in order to obtain the license "to cover employees we do not have and do not intend to have".

The contract was then terminated for default and on the same date reprocured to another contractor. The Interior Board of Contract Appeals decided the default termination was justified and also because the original contract was for an estimated quantity which could be up to plus or minus 20 percent then the original contractor was held liable for all the excess procurement costs. This included the seven percent over-run on trees that the second contractor planted.

Refer to: Thumpers Reforestation, IBCA-1576-5-82

Note: The A.R.C. has received complaints from members that this same BLM District which can award contracts under $10,000 on the District level has subsequently been doing so to unlicensed contractors and allowing these contractors to complete the projects without ever obtaining the necessary license.

WILDERNESS CONTROVERSY

Last year a coalition of Democratic representatives led by Jim Weaver introduced a resolution (HR 1149) into Congress that recommended increasing Oregon's wilderness holdings over 1,000,000 additional acres. Many of the proposed additions included some of this nation's most productive timberlands. In a March, 1983 Board meeting the ARC Board of Directors unanimously supported a resolution to oppose HR 1149 as being excessive and detrimental to the economy of the state. A result of this resolution was that a letter was sent to Senators Hatfield and Packwood informing them of our position. Both Senators replied immediately, with Senator Hatfield soliciting more specific comments for his field hearings. The following statement was drafted following last August's annual ARC meeting and was presented to Senator Hatfield's committee on August 25.

HEARING ON HR 1149 BEFORE THE SUB-COMMITTEE ON PUBLIC LANDS AND RESERVED WATER OF THE SENATE ENERGY AND NATURAL RESOURCES COMMITTEE

STATEMENT OF
ASSOCIATED REFORESTATION CONTRACTORS, INC.
Salem, Oregon
August 25, 1983

Dear Senator Hatfield
and Members of the Committee:

In a letter to your office dated March 30, 1983 I outlined the Associated Reforestation Contractors, Inc. unanimous opposition to HR 1149 as well as some of the points of discussion concerning that bill. In your April 25th reply you requested "any particular comments" that we might have. Copies of those letters are included with this statement to provide a background to the following commentary:

During our annual membership meeting held this year on August 19 at the Agate Beach Hilton, I made a formal presentation concerning the wilderness issue in general, our board's position on HR 1149 and the basic reasoning involved in adopting that position. Following an outline of six other wilderness proposals, the floor was opened to discussion concerning any specific comments our organization might make to this hearing. The following points were discussed without attracting dissenting viewpoints:

> Any proposed additions to existing wilderness should allow for labor intensive management. Specific mention was made for fire control, trail maintenance and insect control through salvage harvesting.
>
> The proponents of additional wilderness form a very small, vocal minority. Actual users of wilderness form a small minority of the proponents. It was felt that these individuals should not be allowed an influence exceeding their numbers and that greater consideration should be given to the original agreements by which Oregon lands were first placed under Federal management; in particular the issues of community stability and revenue sharing.
>
> In any instance, the issue should be settled as quickly as practical.

Points of discussion that attracted *opposing* or *questioning* viewpoints were:

> Commercial grade timberland should not be considered "wilderness," but rather as Oregon's most valuable crop land. It was argued that Oregonian workers and property tax payers would be, in effect, subsidizing "wilderness" involving crop lands through job loss and harvest profit loss.
>
> Eastern and Southern States should be expected to provide similar amounts of wilderness to those already existing in Oregon before expecting this state to further encourage this unpopular form of land management.

A conclusion was reached that the ARC, through time and financial limitations, should not develop its own proposals, but rather, should support the existing proposals that most nearly paralleled the views of our members.

During the evening business meeting, following the election of new officers and board members and the formation of trade committees, the position of the new board concerning the wilderness issue was discussed. It was decided to support one of the following proposals:

1. The Regional Forester's Proposal
2. The January, 1979 Final Environmental Statement
3. The Carter Administration Proposal
4. Governor Atiyeh's 1979 Proposal
5. Senator Hatfield's S. 2031 Proposal
6. The AOL January, 1982 Proposal
7. HR 1149

Serious consideration was given only to the Atiyeh, Hatfield and AOL positions. It quickly became apparent that the majority of those present considered the AOL proposal as best representing the concerns of our organization.

As a result, the Board of the Associated Reforestation Contractors, Inc. voted unanimously to enter a statement to this hearing fully supporting the Associated Oregon Loggers 1982 Wilderness Proposal.

Respectfully Submitted By,
Bob Zybach, Vice-President
Associated Reforestation Contractors

SITE PREP

By Bruce Fraser

SITE PREPARATION

Site prep includes pre- and post-logging slashing, constructing fire trails with hand tools such as shovels and pulaskis, helicopter applications of herbicides and dessicants, helicopter and hand ignition of slash piles and broadcast burns, laying primer cord and jelly gas and mopping up controlled burns and "slop-overs" with chain saws, hand tools, hoses and sprinkler systems as well as aerial applications of fire retardants. Also included are plowing fields with conventional farming equipment, scarification (removing surface and subsurface biomass with a caterpillar or high lead system), yarding unmerchantible materials (YUMing), snag falling, and hand application of herbicides.

Rodney Slattum, Crew Foreman
Phoenix Reforestation

A vast amount of timberland in Lincoln County was logged off without any plan for subsequent reforesting after the harvest. This is why so much of the forest here is either dominated by weed trees, or has just been re-established as conifer plantations in recent years. To quote an elderly pioneer logger from Harlan, "We just creamed the hell out of her and it shows today."

Tom Swinford

To convert their trash lands around their Toledo Mills into good tree farms both Georgia Pacific and Publishers have conducted on-going programs to maximize the future yield from their Lincoln County holdings. Both of these timber companies hire local reforestation crews to do their site preparation. Often the same crews will plant these sites the following winter. For the past several years G.P. has contracted with Phoenix Reforestation of Eddyville, and Publishers has worked with Dean Pihlstrom Inc. of Newport.

Bob Zybach is the main man at Phoenix Reforestation Inc. The definitive sub-title at the top of this article is taken from his paper entitled Reforestation Practices, used last year as an aid by the Northwest Ratings Bureau to establish Oregon's Worker's Compensation class codes for Reforestation. Over the past eight years Bob's cutting crews have slashed several thousand acres of weed trees in the Coast Range.

Phoenix Workers know quite well what good slashing can mean to the planting operations to follow. On one long term contract in 1975 they weren't required to cut vine maple, but merely to drop the dominant, full sized Alder on top of it. There was no burn scheduled for this unit on which heavy pulp logs were yarded with a D-8 Cat. Though the Cat Skinner was good, the vine maple could neither be torn out with the blade, or crushed by the tracks of the D-8, as was hoped. The vine maple was a tremendous obstacle for the planters this winter, and has badly hurt the growth of the new Forest there. That experience brought home the value of the thorough clear-cutting that is a crew trademark on all their slashing jobs since then. Phoenix does burning too so they also readily appreciate the importance of good slashing from the standpoint of creating good, hot fires, and the safe, high-production planting sites that are created as a result.

Steve Huguley

Dean Pihlstrom crews work on some slashing contracts, but coastal residents recognize them far more for their professionalism in broadcast burning. Dean's business card features Control Burning as one of the services his company provides. The high columns of smoke that fill the summer skies over Newport bear this out. In the 5 years that Pihlstrom crews have done control burning in Lincoln County they've put over 10,000 acres in the air.

This series of pictures shows a helicopter ignition of a broadcast burn. They were taken by Dean at a burn conducted jointly by G.P. and Publishers on a tract where they are adjoining landowners. This was a smooth collaboration involving both Bob's and Dean's crews deployed over the units of both timber companies.

Dean has never had a major accident on his burning crews. One reason for this is probably because of the vast amount of experience the men on this crew have. over the course of the season he's built a 20 man burning crew from the best of the 100 workers he employed during this year's planting. His crew list included five quality foremen. Two of the foremen do the snag falling that precedes many burns. One advantage to having so much experience is that the snag cutters can work with the fire trail crew to even out fuel concentrations on the perimeters of the unit, thus helping to protect any standing timber bordering the burn from fire damage.

Although the burning Dean is currently involved with is done entirely by helicopter ignition, he'll quickly point out the circumstances in which hand ignition by drip torch, or mass ignition with primer cord can be preferable. On every unit decisions must be made from evaluations of unit size, steepness of slopes, fuel concentrations, moisture, the type of surrounding forest, fire trail, road access, wind conditions, personnel, water resources and tool supplies.

Dean believes in long range planning in the logistics of deploying men and equipment for containment and mopping up operations. He states his professional philosophy toward broadcast burning as being simply . . "though fire must be regarded with caution it is emminently controllable."

A Controlled Burn

As of August 31st this had been a slow year for slash burning operations at Publishers Paper Company's Toledo Timber Division. It's been a wet Summer on the Oregon Coast. As a result the burning of nearly 1,400 acres of slashed and logged off Publishers' ground had to be postponed repeatedly as fickle rain clouds doused the coastal hills time and again. When the weather forecasters confidently predicted a short streak of true summer weather it obviously meant an opportunity to do some catching up on the backlog of unfinished site preparation. Publishers Foresters weren't about to waste this chance and scheduled their work accordingly.

On Tuesday the 31st the main landing on a Publishers burn site between Logsden and Nashville was crowded with visitors who didn't usually attend these functions, including myself. The main burning crew of Dean Pihlstrom Inc. was there to do their job, which on this day looked to be about 400 acres worth of mop-up firefighting.

Besides Dean and his crew there was an entourage of Publishers' Forestry personnel headed by their boss Shannon White. G.P.'s Eric Bundy was there along with an assortment of observers from the Oregon State Forestry Department. The occasion that brought them there was the testing of some foaming chemicals, and the demonstration of some new foam spraying equipment by its inventor, Al Crosby.

From watching the videotape I I took of the tests I'd have to say that the sprayer can pay for itself in spot situations, especially in containing heavy fuel concentrations and protecting high priority timber. It's a little awkward in its current form to be very effective in inaccessible areas, but with some modifications it's conceivable that foam spraying might grow to be as worthwhile in forest fire prevention as it has been in all fire fighting.

Slash burns on State or privately owned ground require a permit from State Forestry. They base their decisions largely on the Oregon guidelines for smoke management. The burn this day was somewhat of a determining factor in obtaining an unrestricted permit for the following day. Publishers Forester Shannon White and the State Forestry Inspector (Toledo Dist.) in charge of burning permits were busily discussing where to light the fire today to best shape proceeding operations. I asked Shannon how it looked for being issued that permit.

"I don't know yet," he replied, "It's still too early to tell." Then motioning to the nearby inspector he added jovially, "That's up to this gentleman right here."

"Now he calls me a gentleman," the State inspector laughed.

Any slash burn requires some degree of waiting. While we watched the helicopter G.P. Forester Eric Bundy answered some of my questions about helicopter fire lighting. He described the copter's flame dripper, explaining how it ignited a mixture of regular gas and Aluma-Gel with a flint device on the spout of the dripper's torch. I asked him what difference the experience level of the pilot could mean to burning jobs.

"A lot," Eric said. "A Pilot who hasn't worked or seen many burns can hit the wrong spots and cause an uneven burn. These guys have to be able to fly in dense smoke and still hit their targets. It often means relying solely on instrument readings for accurate ignition in zero visibility."

We discussed how G.P. was benefiting from site preparation, and specifically how Phoenix Reforestation's crew was doing on G.P. ground this summer.

"They're doing pretty good despite this wet summer. It's great how some of the guys who've been around awhile are putting out. I think it's partly because they work all year and can see how much good they're doing for themselves as well as us.

On the last few passes of the copter I got in position to shoot some videotape and could see how the pilot was drawing his flame lines and getting the most out of each load of igniting fuels. It was a long stretch of units being burned so that meant a long fly back to fuel up the drip torch. In all it took slightly over 2 and a half hours to ignite this burn. Since the State permit dictated that they start lighting around noon it wasn't until about 2:30 that the bulk of Dean's crew came up in the fire trucks assigned to this section. I rode with Foreman Dave Maier down to the farthest landing where Dean had already dispatched a pumping outfit. Enroute I asked Dave about snag falling when it's done just as preparation for burning.

"I don't fall every snag on the unit," he said. "Since I see all these burns, and the eventual planting too, I know I don't need to drop them all. I just drop the ones I think will help the burn go better. It's safer snag falling and the owner gets all the benefits of a better fire without having to hire a full time faller."

Through the smoke we saw a converted dump truck with a custom built 3,000 gallon water tank filling a portable pool. Crewman Larry Robison was rigging a pump as the pool filled. Dean and other crew men were on the unit scouting for hot spots. When the pump was ready I asked Larry if they'd had any "slop-over" this year.

"Sure, fires jump the lines every so often. If they didn't we wouldn't be here," he said jokingly. "Nothing real serious though. I wish he'd touched off that string of salal and vine maple," Larry said as we watched the fire cleaning the slope above a flat valley where the brush hadn't been ignited. "That will be tough to plant. It's too bad because that salal burns real well. Most of that steep ground is going to be a lot easier."

Dean hiked out of the unit and prescribed the required number of T's, Y's, nozzles, and lengths of hose. The men fixed their bandanas over their noses and hiked down into the smoke toting the gear their boss recommended.

Dean started the pump and being sure that the water was working in this area shifted his concentration to another section of the burn. His crew is well equipped with radios both in the rigs as well as with hand held walkie-talkies. The radios were calmly busy with a steady exchange of information among the crew. As we rode up to my truck I asked Dean if he saw any reason why they might not get the unrestricted permit tomorrow.

"None," he assured me. "As well cleared off as this is right now I'm pretty sure this will all be cool and blown off by morning. They'll probably get 2 helicopters out here and really let it rip."

As Dean predicted State Forestry did issue extensive burn permits for that Wednesday. Not only did Publishers burn over 600 acres the next day, but G.P. burned 390, and Starkers 143, all in the same general area.

On September 1 there were a lot of complaints from folks in Newport about the pall of smokey haze that loomed in the sky to the North and East. It is unfortunate that the best days of Summer are often the best days for burning, but as I try to explain to people, it's hard to appreciate slash burning if you don't plant trees for a living.

Brush Control

The Forest Service Takes a Ewe Turn

By Bob Zybach

Wanted: 5000 vegetation control experts for 1000 acres of forest plantation release work. Must be 60 days to 3 years of age in sound physical condition and preferably covered from head to hoof with fine, curly hair. 12-16 hour shifts, 7 days a week.

Pay: All the thimbleberry you can eat. No overtime compensation.

Holy Davis-Bacon Act! Is this the additional employment Jim Weaver promised reforestation workers with the banning of herbicides on government forests? Possibly so, according to Gene Klingler, silviculturist for the Alsea Ranger District of the Siuslaw National Forest. In a study partially initiated by concerns expressed by recent immigrants to the central Oregon Coast Range over the safety of herbicide use by timberland managers, researchers came to the conclusion that sheep may be a valuable source of biological control of unwanted vegetation.

Mr. Klingler is using the results of that research to investigate the cost-effectiveness of sheep use on a larger scale. Other districts of the forest service and BLM are following suit.

Investigators Steven Sharrow, Wayne Leininger and Bruce Rhodes, operating on a $100,000 grant given to Oregon State University by the U.S. Department of Agriculture, recently came to the following conclusions after three years of study:

1. Sheep grazing on new conifer plantations may provide a cost-effective means of controlling competing vegetation in high-site areas.

2. Such grazing offers an increased potential for red meat production.

3. Sheep grazing reduces the amount of damage to conifer seedlings by big game animal browsing.

4. Forage for big game animals is improved in areas previously grazed by sheep.

5. Sheep grazing may increase diameter growth on established conifer seedlings.

Although point 5 may appear to be relatively insignificant, it is worthwhile to note that the five study plots used in this project averaged a site index of 170 feet/100 years according to Mr. Klingler. The additional diameter growth noted in the study plots, unusual for such high sites, was probably due to fertilization techniques employed by the sheep during the course of their brush suppression activities.

A similar reasoning is used to rationalize the apparent contradiction in points 3 and 4. It has been established that odors from sheep and goats are sufficient to repel most deer. Additionally, most effective grazing by domesticated animals occurs during the same period (spring) that most damage is done by wildlife. Depending upon rainfall, it is usually 3 days to 6 weeks before deer will return to units browsed by sheep, at which time well fertilized forage more desirable than hardened conifer shoots is then available.

Point 1 seems to have been borne out by the studies undertaken by the above investigators in four of the five areas tested. Point 2 compared weight gain (or loss, in the case of ewes nursing lambs) with that of ewes and lambs raised in the Willamette Valley. Results in both cases seemed very encouraging.

Although the Douglas Fir region of western Oregon and Washington has long been recognized as one of the greatest timber producing areas on earth, and although the relationship between sheep browsing and conifer reproduction in this region goes back nearly a century (F.V. Coville put out USDA Division of Forestry Bulletin No. 16 "Forest Growth and Sheep Grazing" in 1898), it is only with the above study that a systematic review of the literature on the subject has taken place. This is probably because it has only recently been recognized that brush control is a serious problem in seedling establishment in high site areas, or possibly because of a traditional reliance upon herbicides to effect that control.

Private landowners, contractors, foresters and others interested in further investigating the potential for cost-effective simultaneous timber and meat production, should contact the Siuslaw National Forest (headquartered in Corvallis, Oregon) and request the "Sheep Grazing in Douglas-Fir Plantations of Oregon's Coast Range and Its Impact on Big Game Habitat" Final Report 1980-1983 By Rhodes and Sharrow.

"Forage Preferences of Herded Sheep as Related to Brush Control and Seasonal Browsing Damage to Douglas-Fir Regeneration", Submitted October 1, 1983, By Sharrow and Leininger.

For those interested in applying for the positions listed at the beginning of this article it is helpful to be aware of the following: Lambs younger than 60 days will not be considered for work on steep slopes because of their inability to stay with their mothers. Columbia and Rambouillet breeds are preferred because of their strong herding instincts. Angora Goats will not be considered at all.

Note: According to Associate Chief Forester F. Dale Robertson in May, 1984, the Forest Service may not resume spraying in Northwest National Forests before 1986. He also stated that the service has no plans to ask Congress for more money to boost mechanical brush control that cannot be included in funds appropriated for fiscal 1984 and 1985. ♠

YOU COULD LOOK IT UP

By Bob Zybach

New Study and Statistics Examine Issue of Employee Health and Safety Hazards Associated With Site Preparation, Conifer Release in Pacific Northwest.

Over the past few years I have had the pleasure of working with Bruce Fraser on a variety of projects involving the use of herbicides and/or chain saws during the course of reforestation projects. Some of these projects culminated in Bruce's two part series that was printed in this magazine earlier this year. This article is meant to supplement and update Mr. Fraser's work.

The first article examined the question of whether modern herbicide use constituted a serious potential health problem for those of us most consistently and most heavily exposed to them in the commercial forest environment; regeneration foresters, service contractors, and in particular, the employees of both. As an editor of this Quarterly at that time, I had several occasions to discuss with Bruce the apparent lack of meaningful information available concerning accidents and illnesses in the reforestation industry. The primary conclusion that Bruce drew, that chain saw use posed a far greater hazard to the worker than did herbicide use, came largely from his own experience, conversations with contractors, and an apparent lack of herbicide-related compensation insurance claims on file. There appeared to be no literature or reliable data on the topic available.

In the second article Mr. Fraser examined the potential for psychosomatic injuries and criminal violence resulting from the anti-herbicide movement's emotional claims being commercially publicized to an apparently sympathetic audience. Particular attention was given the general public's documented phobia regarding the use of helicopters in the application of herbicides. He also briefly examined the oft-repeated claim that a ban on forest-related use of herbicides would result in additional employment for that portion of the reforestation industry engaged in supplying labor-related services.

Last July Jim Kadera, a forestry writer for the Portland, Oregon "Oregonian" newspaper (the state's largest circulaton daily), wrote a perceptive article concerning the federal government's approach to vegetation control following recent court decisions to effectively ban the use of herbicides on federal lands (over ½ of Oregon) during all phases of reforestation. Mr. Kadera is well known throughout the Pacific Northwest for the quality of his reporting on issues of importance to the fishing and forestry industries, so his article, "Spray injunction creates costly jobs" was of especial interest to me. Some quotes attributed to Dr. Michael Newton of Oregon State University contained in the article were particularly interesting in that they, a) corroborated the conclusions that Bruce Fraser and I had been coming to and, b) they were contradicted in the same article by BLM's Edward Ciliberti and USFS's Randall Perkins. Some of Dr. Newton's concerns were amply illustrated by a photo accompanying the article which featured a teen-ager on Forest Service property operating a chain saw without gloves, hard hat, chaps, or, any other apparent safety equipment.

Mike Newton is that rarity among research foresters; the man actually practices what he preaches (and upon his own timberlands at that). His wife Jane has been active in her own right in attempting to throw light upon many of the fraudulent and misleading claims being promoted through the popular media by self-proclaimed "environmentalists" during the past decade. Dr. Newton is best known for his work with forest herbicides and is internationally recognized as one of the foremost experts on the topic.

As a neighboring tree farmer and fellow member of the Oregon Small Woodlands Association, I have felt fortunate in having occasional access to Dr.Newton's expansive knowledge on the subject of herbicides during the course of local tours and meetings. Upon approaching him concerning the accuracy of the quotes attributed to him by Mr. Kadera I was pleasantly surprised to find that he was in the process of revising and updating a 1981 work completed in collaboration with Frank Dost (a well-known and well-respected Doctor of Toxicology at the same University) for the State of Washington concerning the environmental effects of vegetation management practices on State forest lands. Of particular concern to Dr. Newton was the same problem encountered by Bruce and I; the apparent lack of detailed information available concerning potential health hazards associated with vegetation management techniques using methods other than aerial applications of herbicides.

After consulting with Jim Stauffer, ARC President, and Linda Carbone, ARC Administrator, it was decided to give what information was currently available in the Association files regarding ARC-EBI Group Compensation Plan accident figures to Dr. Newton in an effort to aid his study. The potential use for information of this sort as an aid to loss control is readily apparent and it was felt that a report that was as accurate as possible might ultimately reduce compensation insurance costs by correctly identifying potentially hazardous methods of vegetation control, an increasingly important aspect of reforestation contracting. Additionally, it was decided to attempt to obtain any specific information desired by Dr. Newton, if possible, through that portion of the ARC-EBI contractual agreement dealing with the availability of computerized accident data for loss control purposes.

Unfortunately, EBI was either unable or unwilling to release the information requested despite several attempts on the parts of Linda Carbone, myself, and Dave O'Donnell, the Group Managing Agent at the time. Subsequently, Sharon Kelly (the present ARC agent) was able to obtain the services of her son, Zane Kelly, an employee of the Accident Prevention Division of the State of Oregon Workers' Compensation Department, in obtaining the statistics desired by Dr. Newton.

By the time that Zane could obtain 2 comprehensive summaries of the statistics concerning the 1689 lost work time accidents possibly attributable to the reforestation industry in Oregon occurring between January 1, 1980 and June 30, 1984, the deadline for Newton and Dost's report had arrived. Their September 24, 1984 Report **Biological and Physical Effects of FOREST VEGETATION MANAGEMENT** was submitted 3 years to the day after their earlier draft **Environmental Effects of Vegetation Management Practices on DNR Forest Lands.** It is currently being prepared in book form and should prove to be the most authoritative text available on the topic when it reaches print. The methodology developed to analyze potential health problems due to vegetation management practices is such that alternative information would be difficult to develop without using a similar format. And, although existing data is acknowledged as being very general and incomplete so far as vegetation control methods (other than the aerial application of herbicides) are concerned, it is seemingly unlikely that conclusions other than those currently held by Dr. Dost and Dr. Newton will be reached.

Of the nearly 400 pages of the report, only about 50 pages are devoted to "Non-Chemical Control Methods", with a large portion of that section concerned with the toxins associated with slash burns. The report effectively demonstrates the reasoning developed to arrive at the stated conclusions and supports that reasoning with as much documentation as currently exists. For biological control of unwanted brush species the obvious answer is to establish conifers as quickly as possible. To reduce the toxins associated with slash burning the not-so-obvious answer is to first spray the vegetation (fuels) with appropriate herbicides. The resulting dessication results in higher combustion temperatures which produce fewer toxins

A smoke-cloud from a Georgia-Pacific broadcast burn looms over the rural community of Logsden, Oregon.

with greater convection, thereby dispersing the lesser amount of poisons over a greater area. In numbers that would work out to something like this: 10 to 200 tons of material is ignited during a broadcast burn. About 25% to 50% of this fuel actually burns. About 1 or 2 pounds of herbicide is applied several weeks before the burn, about 1/2 of which naturally breaks down. Upon combustion the herbicide creates toxins very similar to those created by the slash. For each ton of slash (or herbicide) burned about 14 pounds of particulates are created. Due to drier fuels, initial combustion takes place at greater temperatures causing the particulates to be ejected higher into the atmosphere, resulting in a lesser exposure to the population.

(The problem of smoke toxicity is not new to Oregon. It is, however, currently at the lowest level in recorded history, principally due to modern burning techniques that often rely heavily on the use of herbicides and helicopters. From the time of the earliest settlements in the Willamette Valley in the early 1800's until the first fire controls were established 70 or 80 years ago, smoke was an annual problem West of the Cascades and was often blamed for causing illnesses and discomfort. There are several recorded instances during this era in which the sun could not shine through the smoke and lights were required to eat noon-day meals during portions of the August-October smoke season! Some years the smoke would last for several weeks at a time. Since the early 1900's until about 1950 smoke was a sporadic problem, although often more severe than in the previous century, largely due to greater logging slash accumulations and cooperative fire suppression measures. The past 30 years has seen the least amount of forest smoke affecting local populations in Oregon's history, mostly due to better tree utilization and regeneration management procedures.)

The methodology used by Newton and Dost considers two basic aspects: 1) the likelihood of human health problems associated with the various methods of vegetation control, and 2) the effects on site productivity and non-targeted plants and animals. The summary of conclusions regarding their report is very interesting; "Keeping labor-intensive practices to an absolute minimum", is the first conclusion listed, and perhaps the most controversial so far as our industry is concerned. The book generated by this report should be read and referred to by all serious contractors and foresters dealing with site preparation and conifer release.

Although Mr. Kelly's statistics couldn't be gathered in time for the report, they will probably be incorporated into the

A crew of Phoenix Reforestation workers prepares to re-fuel a drip torch for a slash burn in progress.

book. This information will undoubtedly support, rather than contradict, the report's conclusions. The statistics are limited due to several factors, not the least of which is their very general nature. For instance, although we can determine whether an injury to an individual was caused by a "fall down on" the ground as opposed to a "fall down to" the ground, (whatever that difference might be), we still cannot determine what the person was doing at the time of the fall or what caused him to fall! Likewise, if a man were cut by a chain saw, we can't tell if he was cutting firewood, mechanically releasing conifers, snag falling, or pre-commercial thinning. This lack of specific job and accident descriptions is very serious so far as loss control is concerned. If we are going to be able to use statistics at all for loss control, then the statistics are going to have to be more precise. Otherwise, they are nearly useless for this purpose.

Of the 1689 accidents **possibly** attributed to reforestation work (many of the accidents are nursery related, Christmas tree related, farm related; others are missing, or misclassified), I was able to determine that 626 were **possibly** related to chain saw use, with about 2/3 of this number **probably** chain saw related. In other words, about 25% of all reforestation accidents seem to be chain saw related. How many of these are related to vegetation control is currently impossible to determine, but my guess is that most of them are. The statistics list 0% of the injured as reforestation workers, less than 75% as "miscellaneous laborers", whatever that means. Another serious problem with the statistics is that they don't list the owner of the land upon which the accident occurred. Thus, unless there is a change in data gathering procedures, it will be very difficult to determine the exact effect the federal government's ban on herbicides will have upon the health of reforestation workers. Remember, these are lost work time statistics and, as such, fall squarely between all injuries (including the most incidental) and fatalities.

So far as herbicides related injuries are concerned, there appeared to be two cases (less than 1/10 of 1%), and one case of smoke inhalation. By comparison, there were 17 cases of natural plant poisoning (probably poison oak)! It would appear that the best way available to reduce poisoning in the woods is to treat poisonous plants with effective herbicides prior to physically entering infested areas.

The 7182 logging injury statistics available for the same 1980-94 time period generally support the same findings with the same limitations of interpretation. All of the reports are publicly available through the APD.

A significant help in analyzing these statistics would be provided by a) listing the landowner, b) listing a standardized job description, c) listing a standardized accident description, d) listing lost work day totals, and 3) standardizing accident "sources". Until such an effort is made to isolate accident causes, hazardous jobs, and work locations, the statistics being gathered and analyzed by APD will serve only the most general of purposes and can only prove to be ineffective for specific loss control needs. In the meantime we are left with the probable assumption that alternative methods of brush control (other than aerial applications of herbicides) are probably far more injurious to health, far less effective, and usually a lot more expensive. ♠

It is the ARC Quarterly editor's policy to publish summaries of reforestation decisions of appeals to the various Federal Courts and Agency Appeal Boards.

Anyone desiring copies of a complete decision should mail $5 ($10 non-members) for each decision requested to the ARC address. Please include with your request both the decision number and the name on the decision. Example: Manuel Galan, AGBCA No. 83-110-1. Allow two to three weeks for delivery.

(Donley continued)

It appears the decision was rendered basically on one criteria, "The contractor's duty to inquire". Donley Contractors didn't realize they had a duty to inquire and they were uncomfortable giving testimony on verbal conversations that became their word against the contracting officer's word, therefore they did not enter verbal conversations as evidence in the appeal. If that evidence had been introduced, would verbal inquiry have fulfilled the duty of the contractor or is it his duty to produce written inquiry when the terms of a contract are not adequately defined? Is it a contractor's duty to monitor the contracting officer's performance? That would certainly make it especially hard for contractors new to the business and if all contractors are not expected to monitor contracts sent out by the contracting officer it certainly discriminates against those who have the duty.

Carelessness of Contracting Officers to adequately perform their duties renders financial hardship on contractors and when they are upheld by the Board of Contract Appeals by passing the responsibility on to the contractor to monitor the contracting officer's performance it fosters distrust and suspicion.

I would like to refer to some statements made by the C.O. through his attorney to the Board of Contract Appeals. He states "If contractor in fact noted the absence of Exhibit 1, and was confused thereby, he had no right to rely on a unsupported conclusion but was in fact on a duty of inquiry". He forgot to mention a telephone conversation refered to earlier in this article. He also states, "Contractor of his own volition accepted additional work under Bid Item 3, by Contract Modification No. 1, at the bid price of $125.00 with full knowledge that 24" x 24" scalps were being required, thus indicating that the contract price was reasonable". Here again he fails to mention the date on contract modification No. 1 which was April 22, 1982. The dispute over scalp size didn't arise until May 31, 1982. How could we have had full knowledge that a 24" x 24" scalp was required?

In the letter that Donley Contractors received from the Board of Contract Appeals they state, "The contractor had just completed work on an adjoing Ranger District where hand scalping was required. This hand scalping was done with 24" x 24" scalps. The contractor had completed contracts in previous years on this forest where hand scalping was required at 24" x 24" spacing and he should have been fully aware of the size of scalps required". The adjoinging district referred to was the Dubois Ranger District whre the scalp size was described distinctly. It is unclear to me why specifications on the Dubois District should apply to the Island Park District.

The type of reasoning used by the C.O. and the Board of Contract Appeals makes me doubt their ability to make just decisions.

Douglas Donley

Native Forests of the Northwest, 1788-1856

American Indians, Cultural Fire and Wildlife Habitat

By Bob Zybach

A surprising number of current management issues in forestry are centered upon the idea of encouraging "native" or "natural" landscapes of plants and animals. Forest and wildlife managers—including small woodland owners—are being forced to deal with an increasing number of laws and proposals regarding such concerns as old-growth retention, snag recruitment, endangered species protection, riparian zone management regulations, logging road construction and National Park Service wildfire policies. Because the stated intent of much of this legislation is to maintain viable populations of select "native" species of plants and animals in their natural (or "wild") state, managers are becoming more dependent upon basic understandings of local forest and fire histories for their interpretations of required and intended practices.

The primary question, then, is: What plants and animals are natural or native to an area? To readers of this magazine, the question becomes more specific: Which current plant and animal populations in the Pacific Northwest are actually native to the forests and basins of this region? A secondary question—and the underlying point of this article— is: How many of these species were dependent upon Indian burning practices for the maintenance of their required or preferred environment?

My dictionary says that to be considered a native of an area simply requires one to be born there. The same source defines natural as "not artificial;" "normal;" and "untouched by man." The federal Wilderness Act of 1964 defines areas of "natural conditions" as "generally appear[ing] to have been affected primarily by the forces of nature" where "the earth and its community of life are untrammeled by man, where man himself is a visitor who does not remain."

On the other hand, according to the 1963 Leopold Commission report, the "primary goal" of the National Park Service should be to "conserve the scenery and the natural and historic objects and the wildlife therein" by maintaining or—when "necessary"—recreating "the biotic associations within each park . . . as nearly as possible in the condition that prevailed when the area was first visited by the white men."

The contradictions in these definitions (avoidance vs. active management) are generally resolved by defining native wildlife as those local plants and animals in natural conditions of existence at the time of white settlement; a time between 1812 and 1893 for most areas of the Northwest. As the quotations used later in this article indicate, there are few (if any) forested

Editor's Note: All quotations are as originally written, with spelling and style unaltered.

Costume of a Kalapuyan Indian

This man was sketched near present-day Monroe, Ore., by an artist with the Wilkes Expedition of 1841. The Willamette Valley of that time was largely inhabitated by a few hundred Kalapuyan-speaking people that regularly burned the areas in which they lived. This practice created a beautiful and productive landscape of grassy oak savannah and flowering prairie that was much favored by the first American settlers to arrive in Oregon.

The following comments are from a talk by Dennis Martinez, a Pima Indian working with the Ten Tribes Intertribal Wilderness Council of southern Oregon and northern California. The text of the complete talk is titled "Land & Culture" and appears in the winter 1993 issue of *Winds of Change: American Indian Education and Opportunity.*

In the old days there were vast areas where the Indian women burned in the morning in late summer and early fall. These areas were like gardens—kind of like the Garden of Eden that was described by the early pioneer people who came from Europe to that land.

The landscape that people saw when they came from Europe was a landscape that was literally an expression of the culture of Indian people. When we talk about restoring the culture of the ten tribes I work with, we're talking about restoring the land as part of restoring the culture. There is absolutely no separation between the way the landscape looked in pre-contact times, the species composition and the structure of that forest, or that prairie, and the cultural needs and expressions of the Indian people. The land is an expression of the culture, as much as the arts and crafts and ceremonies.

We need to bring back all the plants that were used for food for baskets and for medicine. And we need to involve the children in the process.

areas in the Pacific Northwest in which "man himself [was] a visitor who [did] not remain" at the times of the "first visit[ations] by the white men."

People had settled throughout most of this region during the past 10,000 or 12,000 years; they subsisted by some combination of hunting, fishing, cooking and managing vegetation. Management methods included such activities as firewood gathering, vegetable harvesting, tobacco growing, and periodically burning portions of the landscape for hunting, gathering or personal safety purposes. At the times of the actual arrivals of "the first white male visitors," most local plant and animal populations had already been further altered by contacts with exotic (and fatal) human diseases, metal tools, guns and gunpowder, horses and mules.

Any comprehensive attempt to describe natural conditions or identify native plant and animal populations in the Northwest would require several book length manuscripts. The following selection of brief descriptions and quotations is offered primarily as a "first-step" by which one can begin the process of identifying local species of personal or legislative concern. The quotes were selected on the basis of the time, accuracy and nature of the observation. They are arranged geographically; from the continental divide to the Pacific Coast.

Rocky Mountains. The Rocky Mountain forests, peoples and wildlife of southern Idaho in 1805, as first described by Meriwether Lewis and William Clark, had already been significantly shaped and changed by European influences. Shoshonee men of that time could purchase infant girls as future wives for themselves or their sons, "the compensation given in such cases usually consists of horses or mules." A man that owned only two horses was considered "unfortunately poor," and the tribe's stock "may be very safely estimated at seven hundred of which they are perhaps about 40 coalts and half that number of mules."

Lewis and Clark, the "first white men" to view and describe the Columbian basin, had entered a land in which "people have suffered much by the small pox which is known to be imported;" in which people regularly traded for mules ("generally the finest I ever saw without any comparison") with "Spaniards by the way of the yellowstone river in 10 days;" and where some tribes were forced to "remain in the interior of these mountains at least two thirds of the year" because they had become:

easy prey to their bloodthirsty neighbours to the East of them, who being in possession of fire arms hunt them up and murder them without rispect to sex or age and plunder them of their horses on all occasions.

Cascade Mountains. There are few contemporary accounts of Indian activities in the Cascade ranges. That is mostly because the climate of these mountains (depending upon elevation) are inhospitable to people during much or all of the year. In Oregon, written records of crossing the Cascades begin with beaver hunters in 1826. In Washington, the first good records are from botanist William Brackenridge's May 1841 crossing near Mount Ranier. The following excerpt from his journal illustrates how ladder fuels can develop in wet conifer forests. It also illustrates the ease with which Indian hunting or cooking fires could spread to unoccupied forested areas:

We now pitched our Tent and there being no water near we melted snow to procure Water for Tea. As the Cooking of the Supper was going on, some one set fire to the moss covered trees, and in one minute all around was one glaring mass of flame. Now was a confusion and bustle—to get all of our luggage to windward, & in the midst of the conflagration Dr. P. came near loosing the tails of his Coat by a brand of fire that had accidentally fallen into his pocket.

Modern researchers have used tree rings to help overcome the lack of early written records for the Cascades. Tree rings from old-growth pine on the eastern slope of these mountains show that fire was used by people frequently, burning through some large forests at 5- to 10-year intervals. As a result, early settlers often found a grassy, brush-free environment with huge, wide spaced trees in which they could ride their horses and graze livestock.

Westside forests were far different; Douglas-fir and hemlock instead of pine; wetter, steeper and far brushier. With the exception of scattered high elevation meadows and huckleberry patches, there was little obvious indication of human fire use. Prehistoric fires did seem to occur in the westside forests more frequently than they do now, however. One tree ring study that spanned 800 years in the central Oregon Cascades suggested that the fire cycle for the study area was only 78 years before settlement (1830) and, beginning with fire suppression activities in 1910, has dramatically increased to its current rotation of 587 years!

Interior Valleys. The Rogue, Umpqua and Willamette valleys in western Oregon and the Puget Lowlands of western Washington were occupied by prehistoric peoples that traditionally burned millions of acres of land annually. The result was a nearly contiguous series of great prairies and oak savannahs that extended nearly the entire length of the Cascade Mountains in the western United States. When early pioneers headed westward (1835-1848), it was free land in these grassy valleys that they were seeking.

The earliest accounts of prairie burning west of the Cascades dates from 1826, during which year botanist David Douglas traveled with a beaver hunting expedition from Fort Vancouver to the Umpqua Valley. By 1832, most of the Kalapuyan Indians that Douglas had encountered on his journey had been killed by infectious diseases. Hudson's Bay Company journals from 1825-1834 (most notably those of Alexander R. McLeod, Peter S. Ogden and John Work) contain a substantial amount of detailed infor-

(continued on page 31)

Native Forests

(continued from page 15)

mation that complements and corroborates Douglas' accounts.

By 1841, only a few Kalapuyans still lived in western Oregon, but they were still burning and maintaining most of the prairies, wetlands and oak savannah that had provided their families and their ancestors with berries, seed, nuts, bulbs, weaving fibers, fuel, construction materials and animals. In 1920, George Riddle described his November 1851 impressions of the Umpqua Valley home of these people:

At that time, Cow Creek valley looked like a great wheat field. The Indians, according to their custom, had burned the grass during the summer, and early rains had caused a luxuriant crop of grass on which our immigrant cattle were fat by Christmas time . . . Fortuneately in our case, the land was ready for the plow. There was no grubbing to do. In all the low valleys of the Umpqua, there was little undergrowth, the annual fires set by the Indians preventing young growth of timber.

In the same series of newspaper articles, Riddle also described another Indian use of fire:

The massacre of the Indians on Butte Creek occurred on the morning of the 7th of October, 1855. On the 9th and 10th, the country between Gold Hill and Galesville on upper Cow Creek, a distance of 50 miles, was in a blaze.

Oregon and Washington Coast Ranges. The conifer forests of these mountains are some of the fastest growing in the world. They are bounded by the coastal redwoods of California to the south of the Straits of Juan de Fuca to the north. This area includes the largest Douglas-fir, western hemlock, Sitka spruce, western redcedar, alder and bigleaf maples ever measured. As might be expected, these giant forests have also produced the fuels for some of the largest and hottest forest fires in history. The Tillamook Fire of 1933, for example, burned over 220,000 acres in a single August day, creating a mushroom cloud 40 miles wide that extended eight miles into the atmosphere. This fire, as with the other historical (1849-1933) "Great Fires" of the Oregon Coast Range, is thought to have a human-caused origin.

Available accounts of the indigenous human populations that lived along these coasts begin with a daily journal kept in 1788 by Robert Haswell, an officer aboard Robert Gray's fur trading sloop, the *Washington*. Traveling northward from California in nearly constant view of the shoreline, Haswell noted that "this Countrey must be thickly inhabited by the many fires we saw in the night and collums of smoak we would see in the day time," and "a delightful countrey thickly inhabited and cloathed with woods and verdure with maney charming streems of water gushing from the vallies." By the time the crew had reached the Salmon River, along the central Oregon Coast, Haswell noted that the Indians had "both iron and stone knives" and that "two or three of our visitors were much pitted with the small pox;" both certain signs of local European contact and influence.

The settlements and landscapes described by Haswell were also noted by other coastal ship traders in the 1780s and 1790s. The first records of extensive overland explorations along the Oregon coast began in 1826. During those times, local Indian settlements were found near the mouths and rivers of nearly all of the rivers and major creeks that empty into the ocean. Grass prairies and "brakes" (brackenfern prairies), created and maintained by human fires, surrounded most of the Indian settlements and provided habitat for a wide variety of plants and animals.

Writing about conditions in the Tillamook Bay area in 1856, pioneer Warren Vaughn noted:

At that time there was not a bush or tree to be seen on all those hills, for the Indians kept it burned over every spring, but when the whites came, they stopped the fires for it destroyed the grass and then the young spruces sprang up and grew as we now see them.

Summary. The native forest plants and animals of the Northwest are currently being favored by a number of important laws and regulations. Determining which plants and animals should be considered native, however, is a difficult process. Much (if not most) of the local environment has been directly influenced by human activities for thousands of years, making wildlife populations at least partially dependent upon the actions of local people.

Prior to European contact, periodic burning practices were probably the most significant cultural influence on the Northwest environment. Paiute and Kalapuyan tribes, for example, are known to have purposefully burned thousands of acres of prairie at a time, eliminating unwanted tree growth and favoring certain forbs and bunchgrasses.

The first European influences in the region were probably being felt sometime between 1578 and 1776. Human changes to the environment since then include the introductions of domesticated grazing animals, new human diseases, exotic plants, private property laws, metal tools and guns, and huge increases in the number of people, their pets and their visitors.

Current populations of native Northwest plants and animals exist in an environment that has been heavily influenced by the plans and actions of local people, families and communities through time. The historical record tells us that many combinations of these plants and animals have become threatened in the past two centuries because of changes associated with white settlement. It is important that we recognize the endangered status of these wildlife communities and that we understand the reasons they have become threatened. Our common ethical, cultural and natural futures depend heavily upon the forest management plans and practices that we are undertaking today.

Bob Zybach is a graduate student at Oregon State University, Forest Science Department, in Corvallis. Previously, he was an owner in the reforestation business for 20 years.

Forests of the Pacific Northwest: Sustainable Use and Resiliency

By Dr. Bob Zybach
President, NW Maps Co.

In 1805 the Lewis and Clark Expedition crossed the Rocky Mountains and entered the Pacific Northwest. The members encountered a country unknown to science and well populated by native people. The two explorers and their troupe had traveled to a region sculpted over eons by glaciers, floods, landslides, tides and hurricanes. They found a land dominated by giant rivers, great waterfalls, two-mile-high volcanic peaks perpetually covered with snow, and enormous trees; land that had been populated, burned, harvested and cultivated by people for more than 10,000 years.

Much of the land was so bare of wood that at one point expedition members stole boards from a house so the men could cook supper. At other times they scrounged what "fiew" willow sticks they could find along the river's edge to prepare meals. Treeless prairies extended for miles, often with mounds of harvested roots and bulbs dotting their surface.

Much of the land also was forested. In several locations the men described and measured some of the thickest stands of the largest trees any of them would ever see.

Much has changed in the 200 years since Lewis and Clark traveled through the Pacific Northwest. All of the Indian nations the men encountered, and most of those cultures and their combined knowledge, are gone forever. Millions of people have replaced tens (or hundreds) of thousands. Gunpowder, concrete, metal, fossil fuels, electricity and domesticated grazing animals have been introduced into the environment. Thousands of species and varieties of exotic plants, animals and diseases also have been introduced, and many have spread widely throughout urban developments, farms, forests and wild lands. Vast expanses of old-growth trees have been logged and millions of seedlings planted in their place. Many of the changes seem permanent and there is no going back.

Other occurrences over the past 200 years include numerous catastrophic wildfires, floods, major volcanic eruptions, hurricanes, earthquakes, tsunamis, landslides and other forms of landscape-altering geological and weather-related events. Yet, despite all of these dramatic changes in human numbers, technology, culture, weather and geography, most of the forested areas noted by Lewis and Clark remain forested today. Further, the forests seem to be spreading and likely contain greater numbers and greater volumes of most tree species over larger areas of the landscape than at any other time in history.

FORESTS OF THE PACIFIC NORTHWEST

The Pacific Northwest is defined by the Columbia River basin, the Cascade Mountain Range and the Pacific Ocean. The Columbia travels first south and then west, draining the forested western slope of the Rocky Mountains and the yellow pine-dominated eastern slope of the Cascades, including all of Idaho and most of eastern Washington and eastern Oregon. It is among the world's largest and most powerful rivers, cutting directly through the Cascade Mountains to tidewater before entering the Pacific Ocean another hundred miles west. The Columbia was also Lewis and Clark's principal travel route through the region. Their observations in Oregon and Washington were limited to people and environments directly associated with this great river.

The Cascade Mountains are a north-south chain of active volcanoes that stretch the entire length of the Pacific Northwest, from northern California to southern Canada. They form a continuous rain shadow that separates wet western forests of Douglas fir, hemlock and Sitka spruce from dry eastern forests of Ponderosa pine, lodgepole pine and juniper. The volcanoes, including Jefferson, Hood, Adams, Baker and Rainier, are typically more than 10,000 feet in elevation with treeless, glaciated peaks perpetually covered in snow. Mount St. Helens is probably the most recognized Cascades volcano, primarily because of its dramatic and well-documented eruption in 1980. This peak has been active throughout recorded history, with a number of eruptions being noted by journalists and depicted in drawings and paintings in the 1840s and 1850s. The other Cascade peaks are also considered active. Lewis and Clark, for example, described conditions near the mouths of the Sandy and Hood rivers in 1806 that helped confirm a significant eruption of Mount Hood in the early 1790s.

The Pacific Ocean forms the western boundary running parallel with the Cascade Mountains. Between the ocean shoreline and the tree line of the Cascades exists the largest, fastest growing, most productive (and among the most resilient) conifer forests in the world. In the absence of human intervention, nearly 100 percent of this area would become covered in trees and nearly all of those trees would be conifers – mostly Douglas fir, with belts and stands of Sitka spruce, western hemlock, western red cedar, redwood, grand fir and a handful of others. Willows and cottonwoods might follow the rivers, and alder and crabapple might populate seeps, swamps and landslides, but – in the absence of people and excepting the highest elevations on the Olympic Peninsula – conifers likely would cover the entire region west of the Cascades. The vast majority of trees would be Douglas fir, and the greatest number of these would exist in relatively pure even-aged stands.

...despite all of these dramatic changes in human numbers, technology, culture, weather and geography, most of the forested areas noted by Lewis and Clark remain forested today. Further, the forests seem to be spreading and likely contain greater numbers and greater volumes of most tree species over larger areas of the landscape than at any other time in history.

That much of this timber growing area is not covered with conifers, but with pastures, crops, parks, houses, lawns, highways and parking lots is due entirely to the actions of people. At the time of Lewis and Clark, the same argument could be made for the millions of acres of oak savannah, berries, bulbs, prairies, fields and orchards that people had been maintaining with fire for centuries and millennia. For the past 10,000 years or more, forest patterns have been largely established and maintained by people. The primary methods used to achieve this condition have been the near-daily use of forest products and fire. The incremental and cumulative effects of kindling and firewood gathering and use along navigable waters, near fields, homes, campgrounds, and towns by tens of thousands of families over centuries and millennia should not be underestimated. Lewis and Clark's comments regarding a host's wealth and station were based on the lack of firewood, rather than tribal standing, homeownership, hospitality, or the existence of large amounts of dried or smoked salmon.

At the present time there appear to be more trees over a larger area in the Pacific Northwest than at any other time in history. For 200 years, most of the heavily forested areas have remained as forests, despite the great increase in population. Many interior and adjacent grasslands, brakes, berry patches, and other Indian croplands have seeded (or been planted) into conifers. The conifer forest has expanded outward along its edges and inward into its openings during historical time, increasing in numbers and volume and decreasing in average age and size. Oak savannah and other woodlands have all but disappeared during these transitional conditions.

Douglas-fir has spread through the eastern foothills of the Coast Range and filled previously bare interior valleys, balds and ridgelines with young, even-aged trees. Higher elevations of the Cascades and Rockies have seen huckleberry patches, meadows and bear-grass prairies filled with even-aged stands of mountain hemlock, true fir and lodgepole pine; and hundreds of thousands of acres of eastern Oregon and Washington prairies have developed thick stands of (mostly) even-aged juniper, lodgepole pine and larch.

The forests of the Pacific Northwest are not only resilient, they are aggressive.

CATASTROPHIC EVENTS AND FOREST RESILIENCY

How did Pacific Northwest forests become so capable of persistent population patterns and opportunistic expansion in the face of such dramatic human population growth, technological change, and new species competition that has occurred in the past 200 years?

As a forest historian, it is my observation that the Pacific Northwest forest types share long regional histories of (1) large, generally isolated locations, (2) periodic stand replacement events of catastrophic proportions (hundreds of thousands of acres), (3) constant human fire, and (4) nearly perfect growing weather.

The combination of these four elements produced a relatively stable regional pattern of oak woodlands, grasslands and even-aged conifer forests that enjoyed ideal seasonal growth patterns of wet and dry and a periodic pattern of catastrophic wildfires, hurricanes, floods and volcanic eruptions that could kill hundreds of thousands of acres of trees at a time. These occurrences have guaranteed a steady stream of genetic conifer tests for the 10,000 years or more since people introduced daily fire into the environment.

The forests of the Pacific Northwest continue to reflect patterns of catastrophic events that occurred near the end of the last Ice Age. Melting glaciers were replaced by conifer forests that persist to this day. The channeled scablands of eastern Washington became the flat, fertile plains of the Willamette Valley via a series of floods that carved the Columbia Gorge and created its waterfalls between 12,800 and 15,000 years ago. Most of the area, however, was never under glaciers at all, and was not subjected to cataclysmic flooding, volcanic eruptions, or rising sea levels. It was these areas in which the genes of today's forests were most likely kept. When the waters and glaciers receded and the volcanic soils cooled, these were the trees that provided seed for the new forests.

But forests outside the range of such primary landscape clearing and forming events still had to withstand a deadly (if slightly less fearsome) assortment of hurricanes, lightning strikes and fires, insects, landslides and diseases in order to survive and produce seed. Such trees also needed to grow higher, faster and wider than hundreds or thousands of competing siblings of the same age in order to produce seed for any significant period of time. These latter trees could be capable of lasting for centuries, growing nearly 300 feet or more in height, and being capable of reseeding entire stands or groves with their seed following a wildfire, hurricane, or the abandonment of a field or pasture.

To provide an idea as to the magnitude of the "genetic tests" regularly inflicted on forests in the Pacific Northwest, it is only necessary to look at the litany of catastrophic events that has visited the region during the past 200 years. The 1980 eruption of Mount St. Helens is well known. The Columbus Day Storm of 1962, the most destructive windstorm in Northwest history, is another example. Gusts exceeded 140 miles per hour as the hurricane traveled from northern California, the entire length of western Oregon, to southwest Washington in a few hours' time. There was little advance warning and several people were killed. During the course of this event, an estimated 11 billion board feet of commercial timber was blown down.

Catastrophic wildfire is the most consistent type of stand replacement event in the Pacific Northwest. In a single day, August 20, 1933, the Tillamook Fire burned more than 220,000 acres of forest and killed nearly 8 billion board feet of timber while creating its own weather patterns and a mushroom cloud more than 40 miles wide and eight miles high. The "six-year jinx," 1933, 1939, 1945 and 1951 fires, burned even more acreage. A long-term program of fire suppression, salvage logging, tree planting, seeding and rodent control resulted in the "Tillamook Forest." Other catastrophic wildfires include the Nestucca, Yaquina and Coos fires of 1849 to 1868, and the regional complexes that included the Forks and Yacolt fires of western Washington in 1902 and 1910. The 2002 Biscuit Fire covered 500,000 acres in a month's time.

It is too soon to tell how recent catastrophic wildfires in wilderness areas and parks will respond over time. The 1765 Millicoma Fire may provide a possible clue. When Weyerhaeuser foresters cored more than 1,500 trees in a 100,000-acre grid in 1945 and 1946, they were startled to find that more than 90 percent of the trees were the same species, Douglas fir, and were the same age – 180 years old. At least one forester suspected an Indian-caused blaze of perhaps 200,000 acres that was naturally reseeded to Douglas fir by second-growth trees along the eastern boundary of the fire. As Thornton Munger wrote in 1940:

"The paths of the great forest fires of the last century or two are plainly marked by even-aged stands, consisting to the extent of at least 90 percent of Douglas fir (if within the preferred habitat of this tree), regardless of the proportion of Douglas fir in the original fire killed stand."

HUMAN CATASTROPHE AND FOREST EXPANSION

A final, and sad, note should be made of the resiliency of Pacific Northwest forests since the time of Lewis and Clark. Some mention has been made of the roles of human use of fire in establishing the boundaries and structure of those forests. Another catastrophic occurrence, which had a profound effect on the occupation and use of Northwest forests, was the decimation by introduced diseases of regional Indian nations – including whole towns and families – beginning in the 1770s or earlier and lasting into the mid-1830s.

Robert Haswell, a teenage cabin boy, noted the appearance of metal knives and smallpox scars on Oregon Coast Indians in 1788; sure signs of contact with white traders. Nearly 20 years later, on April 3, 1806, Clark would write:

> …back of this house I observe the wreck of 5 house remaining of a very large village…I endeavored to obtain from those people of the situation of their nation, if scattered or what had become of the nativs who must have peopled this great town. an old man…brought forward a woman who was badly marked with the Small Pox and made signs that they all died with the disorder which marked her face, and which she was verry near dieing with when a girl. from the age of this woman this Disruptive disorder I judge must have been about 28 or 30 years past, and about the time the Clatsops inform us that this disorder raged in their towns and distroyed their nation.

Fifty years after Clark's observances, a medical doctor in southwest Oregon noted: "The Kalawatsets and Coos Indians…have at one time been very numerous, the number of [people] and varieties of habitations showing that every stream and nook was once populated."

The decimation of native peoples caused the cessation of widespread burning. Trees immediately began to populate abandoned towns, campgrounds, fields and prairies, quickly creating new even-aged forests throughout the region.

SUMMARY

The vast forests of the Pacific Northwest are among the most productive and resilient in the world. Many species in the region, including Douglas fir, Sitka spruce, western hemlock, western redcedar, red alder, black cottonwood and white oak, have produced the largest specimens and the most voluminous stands and forests of their species ever measured.

These forests have thrived and persisted in the presence of retreating glaciers, repeated cataclysmic flooding, exploding volcanoes, earthquakes, hurricanes and catastrophic wildfires for thousands of years. The forests of the Pacific Northwest have been observed, visited, occupied, burned and harvested by people for more than 10,000 years. There is every indication they will continue to do so for the next 10,000 years. ●

Dr. Bob Zybach can be reached at ZybachB@NWMapsCo.com.

"Pineros": Interview with Bob Zybach by Brinda Sarathy
October 2, 2005, Roseburg, Oregon

Dr. Brinda Sarathy interviewed me at least twice during her Environmental Science PhD research at University of California, Berkeley. The first interview was by telephone and focused on the establishment and growth of the reforestation industry following WW II, including my personal experiences as a professional tree planter in the 1960s and as a successful reforestation contractor in the 1970s and early 1980s. The following transcript is from our second interview. In 2012, Sarathy's work was published in book form as *Pineros*. This is excerpted from the back cover:

The exploitation of Latino workers in many industries, from agriculture and meat packing to textile manufacturing and janitorial services, is well known. By contrast, pineros -- itinerant workers who form the backbone of the forest management labour force on federal land -- toil largely in obscurity . . . Brinda Sarathy investigates how the federal government came to be one of the single largest employers of Latino labour in the Pacific Northwest. She documents pinero wages, working conditions, and benefits in comparison to those of white loggers and tree planters, exposing exploitation that, she argues, is the product of an ongoing history of institutionalized racism, fragmented policy, and intra-ethnic exploitation of the West.

1. [0:00:00] US Forest Service Casual Labor Contracts (1980s)

Brinda Sarathy: The date is October 3rd, 2005 and I'm speaking with Bob Zybach at "While Away Books" in Roseburg, Oregon . . . [crosstalk while adjusting mics] . . . I would like you to tell me a little bit more about the casualization of labor. In our last interview you were talking about how this is a total casualized labor force. In terms of even, say, rural poverty and unemployment, can you just tell me a little bit more about that in terms of work in the woods?

Bob Zybach: Okay, I think by casual labor force, I mean, we're very much like tree planting started. People would go pick up available workers and maybe transport them long distances. Typically, supply their basic needs as far as a dry place to sleep and food, sometimes work gear, and provide the type of work that you could readily train somebody to do in a few hours' time.

The casual laborers were, when I started, were typically people with alcohol problems, young to middle aged white guys, some Indian guys, one black person I can remember in several years . . . I started about '66 in September, and I worked for Tom McCready until January of '70. When I was first starting as a tree planter, that's all I did for Tom. I worked on his crews exclusively.

Brinda: What was the actual physical labor of tree planting? What was the work like?

Bob: Okay [crosstalk]. The physical work of tree planting is it puts people in about the best possible physical shape you can be in. It's way more demanding that athletics, and you're doing it -- or logging -- and you're doing it typically seven or seven and a half hours a day. With our crew where we moved up and down with the elevations and seasons you might do it eight months or nine months in a year . . . So, it's like being in training camp for eight or nine months. There weren't very many people that worked completely through the season.

There's a lot of us that did, but not a whole lot. Then when we started adding other jobs to reforestation, like pre-commercial thinning, or trail building, or things like that, those are typically easier, and you could pace yourself more. When we did broadcast burning or trapping, it was more of a crew type of involvement. So, it's more back to casual labor quality of work.

Brinda: Okay. In terms of the trail building and all, were these on public land?

Bob: We did very little on public land.

Brinda: Right, you were saying that.

Bob: I did some work on the Sundance Burn in Idaho when I first started . . . They encouraged me to do more work; "they" being the federal government, because we were doing better than the contractor there. Then the contractor went bankrupt on us, and the government, even though they were the ones that asked us to do the additional work, it took two years to pay us fifty cents on the dollar. So we almost went broke by doing high quality work, and extra work . . . we did BLM contracts to start, because we could negotiate those typically, or strategically bid them. We had clusters of BLM work that we could do. The BLM or the Forest Service work was way too restrictive, way too regimented, and it was wasteful. I thought there were a lot of bad habits with it.

Brinda: When you mean restricted, what do you mean?

Bob: It would have requirements that you had to plant trees on 10-foot centers. It would give you low quality seedlings and say, all the root hairs had to be pointed down. They would have somebody with a two-year degree that would go through and dig root hairs, and they would tell you if your root hairs were up or down. Then you'd try to explain geotropism to them, and they wouldn't understand. Then they would dock your pay. They'd say okay, we found 88% good root hairs, or placements so we're giving you 88% pay.

Brinda: Okay, based on a few samples.

Bob: Well this was a statistically valid sample, but the methodology was not statistically valid because they had kids with no experience out there determining quality levels on something that they: one, couldn't have met themselves; but two, it was almost irrelevant. There'd be a J-root study, and they'd say, "then you get surface roots up from J-roots" and we'd say okay. Except we'd planted millions of trees over tens of thousands of acres, and we'd never seen that happen once. Other than that, is that really worth spending a lot of time and effort, slowing us down to dock our pay to give us a problem that doesn't exist and then make us pay for it? So if we had a $2500 job and we only got 90% pay we'd be docked $250. That's what I mean by too restrictive it was just regulated to . . . we could do better work for private industry for about 1/3 the price. So for three or four hundred dollars an acre we could go out and do high quality reforestation work for Georgia Pacific or Weyerhaeuser and then the government would cost 1000 or 1200 dollars for a lesser quality job.

Brinda: As a contractor, would it not be in your best interest to go for the higher pay?

Bob: Well, you lose efficiency, and then you're playing a game with casual labor people. People that would say I can move 20 people into the Rockies, two states away for 32 days and move back because they were running Mexican laborers or because . . . so the contractor specs would be written to . . . There were odd things. There would be a district where somebody says we don't want to hire extra manpower during this period of time, so we're gonna use all of our people that are available and take three months to do a project. Another district would say, we've got such a window of opportunity between the snow, we're gonna hire extra manpower and we're gonna do it in a week. But we're not gonna tell you which days those weeks are in, they could spread over a three-week period, we're gonna wait for ideal conditions. So we've got a crew sitting on standby, which if they're all here illegally that's like their tough luck. If they're all here legally it's like your tough luck, they'll just leave.

It was more important to have steady work that was productive. You felt like you were doing a job rather than meeting a regulation and then we could make a profit. We got paid for what we did, so if we wanted to make more money we just worked harder. The standards we had were based on, interestingly enough, on my quality standard. My quality standards were the common factor through whoever we bid on, and it met all of their quality standards. What I established for our crew as far as Weyerhaeuser, Georgia Pacific, or Boise Cascade or any of the private land owners . . . now I worked for Indian lands too, Makah, Quinault, and we'd occasionally fill in if we got snowed out or caught between contracts, which occasionally would happen, I'd just call up somebody like John Foster and maybe put a crew on one of his jobs for a week or two.

Brinda: Okay, and he was on Federal lands primarily, right?

Bob: Yeah, quite a bit, but he had private lands, and he did tree farms, Christmas tree farms, so he did a wide variety. I focused almost entirely on private tree farms which I did mostly as a favor to the neighbors because I could get cut-rate trees. I'd put our crews in cheap, and they would be lucky if I ever broke even on them, but they kept the crews going during down times. So we never had a down time, we worked year-round I think, for about an eight-year period at one point -- which we were the first crew to work year-round, and then once we did it, it was pretty easy to keep everybody going.

Brinda: Would you say a big part of that ability to work year-round was working on private versus public lands?

Bob: Yes.

Brinda: Could you say a little more about that and why that was?

Bob: Private lands wanted the job done cheap; they didn't want it done to regulation . . . If you said, okay, I was the contractor on a private, I wasn't just some person putting my hat in the ring, I'd say, "we will be here in February and we will complete your job by March 1st."

Brinda: That's what you would say?

Bob: Yeah. For Georgia Pacific, they'd say, "okay, you've got two million trees, can you get it done by April 15?" I'd say, "If we can start by November 15, we can." We went about 20 years and never were late on anything. We had five replants; I called three of them. Other crews would have, sometimes several replants in a day, certainly in a week or in a month's time, that was real normal. Eight years without a [work] day off, people worked . . . and we worked Christmas day and Sundays sometimes.

Brinda: It was primarily what? Replanting work?

Bob: Replanting is when you have planted it and then you've got to replant it. But people do say like that all the time: it's "planting."

Brinda: It's "planting" -- okay.

2. [0:10:50] The Hoedads: Contract Crews & Co-ops

Bob: We did very little replanting. There's two kinds of replanting. One, you do a crappy job of planting, so you've gotta go through and replant in order to get paid and meet job specs, and another kind is you go through and maybe you do a fine job but it's bad nursery stock or rabbits come in or deer come in and they kill off a good share of the seedlings and then you're replanting in areas that have been planted before. When I talked about the Hoedads, we replanted there because it was really crappy work.

Brinda: Right, because they had spaced too . . .

Bob: They'd put tons of trees along the roads and cut banks where it was easy to plant -- apparently, they didn't have an inspector. They had not gone off into the brushy areas; there was nothing. There wasn't any evidence, there was just animal signs. It was what cut corners is, literally, what you do. They hadn't counted reprod.

Brinda: What's reprod?

Bob: Reprod is natural reproduction. You would be in an area and maybe it'd have hemlock and true fir, and Douglas fir, and cedar. Sometimes even they would count alder and things like that; not often.

Brinda: From what it sounds like in my reading, the main groups of the players were contract crews and the co-op crews.

Bob: Yes, for the bulk of the time the industry was doing well, it was that the co-op crews followed the contract crews.

Brinda: They followed the contract crews, okay.

Bob: Yeah, and then you were saying it's best . . . how they were the ones who invented busheling or something?

Brinda: They talked about that they did the quality control, at least in the book, and how they shifted the nature of the game toward being paid by the hour and not stashing trees. Again, based on this one source.

Bob: Yep. They were good at self-promotion. Their main thing was that they had a big, huge business advantage because they didn't pay compensation insurance and quite often, they didn't pay social security or unemployment. But most of these city kids, the quality of their work and their work ethic wasn't very good. But they relied heavily on federal minimum dollar amounts. The federal government says, "you have to pay this much," so they had to bid at those rates. On federal bids another advantage is you had to have a deposit, so they bought land, they had a wealthy person start them up. They used that land to get higher price jobs, ones I couldn't get because I didn't have good enough economic circumstances to bond these; they did. Because they had a ready group of casual laborers, hippie co-operative members, they could get a fair number of people out into a place. They used the federal minimum wage guidelines, which were real high; they helped take them up, they lobbied for that. They were really good politically; they lobbied the state and the Feds for all kinds of advantages.

Brinda: The private land owners didn't have to pay a federal minimum wage?

Bob: Oh, the private land owners had to pay a lot higher than that. Typically, until after everything crashed, then this became a buyer's market. People were busheling and making more than $100 a day in the mid-'50s if they were cutting timber.

Brinda: If they were cutting timber?

Bob: Cutting timber. If they were planting trees at that time, they might make, in the mid-'50s, $25 a day. But the minimum wage was $10 a day . . . It would depend on the work forces. A lot of times they'd make $2 an hour, so like say $15 or $16, but that was still pretty good, especially for guys . . . there'd be a lot of high school kids, just out of high school that would work on local jobs because of the Tillamook Burn, or in local communities. The Forest Service jobs and the ones requiring contractors and better business practices were Bob Snow and Tom McCready and those people.

Brinda: Yeah. Do you know when the Forest Service started contracting out this work? Was it ever in-house when they built the Forest Service or the BLM?

Bob: That's a good question. It used to be in-house for Weyerhaeuser and some of the others, and they used union crews and that was to our advantage because we were way more efficient than the union crews. They ate up way more money than us, mostly in benefits and things, so it made us look good. The Forest Service, that's a good question.

3. [0:15:37] Union Labor & Compensation Insurance

Brinda: So just to back up to the union crews with Weyerhaeuser. Did they always use in-house, were all their crews unionized?

Bob: For, I think significant periods of time they were, where they had to. While I was working with Weyerhaeuser a lot they had to run at least one token union crew. All the union guys that went to plant trees, and they hated it because it's really hard work and the crew was miserable, and they weren't very good at it; we made 'em look bad. There was lot of conflict between loggers and tree planters early on because they saw us as hippies. But then when they'd try to arm wrestle in the bars or things like that or tried to work on our crews, they saw us as good workers. There's a huge difference, because one, they don't want you around the town or their daughters or the bar or wherever they're at; and they don't want to look at you.

Brinda: The loggers?

Bob: Yeah. The other one is, they'll buy you a beer and they're glad to see you. If there's respect, and that took a while to develop. Then it's kinda dissolved like you've been talking about; goes back to the casual labor arc which is the one we're still in now.

Brinda: I know you planted for Georgia Pacific for a long time, but I don't know about Weyerhaeuser. So these in-house employees at Weyerhaeuser were usually loggers, right?

Bob: Or sawmill workers.

Brinda: So they nominally went out to try and plant trees but couldn't really do it.

Bob: They thought they could.

Brinda: They thought they could but then you guys came in is that . . .?

Bob: Yeah, they ran time studies on us and everything else, and they gradually shifted away from the union crews. The union allowed it to happen because they really didn't have good support. The union workers wanted logging jobs which are hydraulics and a lot of sawmill jobs which is under roofs. But the only real thing that compared to what we're doing would be maybe, choker setters or timber fallers, where they're out there, they're in the rain, they're working.

Brinda: They also get paid less right, choker setters?

Bob: Choker setters get paid less but the timber fallers get paid at the top. They work typically, a shorter day. We did when we had saw jobs to do, we'd just pace ourselves to work one hour less a day.

Brinda: In your experience as a contractor, did you ever come across any attempt among the workers to sort of organize against the contractors, or was there a move for we want higher wages, we want to be recognized on par with loggers, this high skill work. Or we want to be unionized because there's a lot of history on loggers unionizing right?

Bob: Well, to some degree. My great-grandparents owned a sawmill. The Wobblies struck against that around World War I or a little before. You're right, the Pacific Northwest, Oregon and Washington, I think were the two most unionized states this side of the Mississippi. In the western United States, and a lot of that had to do with the lumber industry. A lot of it had to do with the industrial lumber industry, and a lot of it had to do with the Wobblies, just getting them unionized.

This was especially a problem in the cedar mills because people would routinely be missing parts of their fingers because these knives would come down. If you lose enough fingers, then you lose your job. You don't have any kind of social welfare to fall back on and there was other problems. They were working 13 and 14-year-old kids in the mills and logging the sites. I know a lot of people that started logging when they're 12 or 14. My great uncle started at that age.

So that was kind of a residual thing, and then in the 50s and 60s there started being a lot of Gyppo loggers and Gyppo sawmill owners and those are independent, family operated, typically three or five or seven employees. They'd buy patches of timber, and they'd sell it, and they would be, of course, non-union, non-benefits, sometimes under the table. But the opportunity would be there to make a lot more money. But then the steady work wasn't there, and the benefits weren't there, so the actual movement would be away from unionization. In my business I actually approached union and tried to unionize my workers, and the union was so suspicious of that approach from me, as an employer, that they backed away from us.

Brinda: Who did?

Bob: The union. We had meetings at my house, and we had meetings with my crew and things.

Brinda: You tried to unionize your workers as an employer, that's very interesting.

Bob: Yeah, well our next step was we had health insurance, and we had compensation insurance, but there were people like Dean Pihlstrom that had these Salmon River Firefighters or the Hoedad's that didn't have compensation insurance that were really gypping the workers. If we didn't have benefits and good wages, it was all going to turn into a casual labor industry done by Mexicans. The idea was to get local kids' meaningful employment with meaningful benefits and to create, as we were fully conscious of doing it, an industry with stability that kids would want to come into. Learn the skills; it takes several years to become a pretty decent tree planter, or it takes several months to develop some pretty bad habits you can't hardly ever break. Chainsaw work is the same thing: several years. To do those basic types of jobs really effectively, particularly within a crew concept, pace yourself accordingly, not be injured, keep working through the year -- and still usually maintain pretty heavy drinking and smoking habits -- requires younger people, in good shape with a good work ethic and good physical capabilities.

Brinda: I'm finding this thing about unionization fascinating, that you attempted that. What union did you approach?

Bob: I've probably got the records on it. Let me think. It would be a labor union, probably AFL-CIO or affiliate I would guess. We weren't . . .

Brinda: And they didn't want anything to do with it?

Bob: They kinda did, but they were very suspicious of "why is a business owner . . .", but the type of business ownership we had was employee stock ownership plan, ESOP. Then the way I ran my business was I had only a percentage of work they did, so I wasn't really in it for a giant income like most of my competitors were. It was perfectly legitimate to be in business to make a profit. I was in business to create a steady, meaningful, good paying employment for me and to use that as a way to segue into some kind of retirement income which I thought would happen in my mid to late 30s. I anticipated owning a tree farm and marketing trees, so I had a real stable work crew, lots of neighbors, some relatives so, legitimately, what was good for them, was good for me.

Brinda: How were you able to then compete with the other contractor's crews if you were providing health benefits and such for your employees?

Bob: We were a lot more efficient. We produced a lot more per day so our gross income per man day or per person working . . . Weyerhaeuser used us as a standard, you could ask other people. I don't know anybody who was more efficient that way. So even though our bid price would be lower, because our production was so much higher our gross income per man day was greater and that allowed me to pay more benefits and higher wages. Then I had zero profit margin, I just was not making much of a profit, we never anticipated having much of a profit. In fact, we would have employee bonuses if there was any profit left over to divvy it up that way.

Brinda: In some senses, you almost were almost like a co-op, but you provided benefits and the worker's compensation.

Bob: We were like a co-op except that we followed the law. The co-ops were, I don't care what they say, they were designed to circumvent the law. There were not cooperative logging businesses, there were not cooperative sawmill businesses, there's not cooperative brush picking businesses, there's not cooperative strawberry picking businesses; why did the Hoedad's have to have a worker's cooperative? That was a way to provide casual labor and that's like collusion basically. We'll all not pay taxes, and we'll create smoke screens and lobby and hire good lawyers. It was an effective strategy for a lot of years.

Brinda: But they finally had to give in and pay . . .?

Bob: Yeah, they went out into the real world and pretty much faded away, I don't know what happened to them. They had pretty good . . . I think they had like 300 employees at a time, so at one point they had to be generating, you would think, $30,000 on a good day. Or you would think $100,000 or $200,000 on the good week to maintain that type of work force. So they were a pretty big business. They defined what the federal standards would be, and then they made those federal standards meet their particular and peculiar combination of attributes, and then they took their pick of the contracts basically, for a lot of years. People like Jerry Rust, and people like that, that's another reason I just stopped. I went, "To hell with it," they can all do that stuff, I don't want to be playing business games. You're either getting overpaid or underpaid from my perspective and that was stressful. I didn't want to make more money than I earned, and I definitely didn't want to keep working for the government and being paid less money and waiting two years on what I earned.

Brinda: In terms of the stuff you were doing on private land, how has that changed over the years and why? I take it you're no longer in tree planting. How did that come about?

Bob: It's basically two ways. In the 80s, in the addition to some of the environmental laws that I was mentioning were kinda debilitating in cost like, there was also a housing depression. There wasn't a demand for products. There was a lot of tree planting going on in Canada and there were some crews that went up there, but then you've got international business problems, and I didn't even like doing interstate too much. They kinda hung on up there. There was a real marked drop-back in federal logging that almost ended; by the early '90s it was gone, but it was like diminishing rapidly. The private lands kinda heated things up in the 80's, but a lot of them then went bankrupt or held back on the reforestation or minimalized their investment because there was such a thin margin to keep things going. A lot of them sold their lands. People like Aaron Jones couldn't buy federal timber, so he bought 120,000 acres of timberland.

Brinda: The Aaron Jones.

Bob: He owns Seneca Sawmill out of Eugene, so whoever's been putting money into reforestation on that. Some people think it's him, he's brand new, he wants to keep cost to a minimum. Then there's these always available, essentially, initially Hoedad crews, they don't have federal work, and they're used to working by the hour and their prices sound cheap compared to my prices. They'll say, "oh we can do that for 15 an hour," then they'd say, "how much for me," I'd say, "I pay my guys from 28 an hour," and they'd say, "wow, why is it so much more expensive with you?" "Well we're doing three times more work, it's cheaper with us." But they've gotta go through the learning curve. They've gotta answer to their people. On the surface it just doesn't look right to charge twice as much to . . . because we're used to busheling, by the acre, by the thousand. There the prices was a way they couldn't compete.

They got a lot of the industry to go over to hourly contracts under the guise that it was cheaper, but the reality was it was easier to sell. "Look at this cut rate we're getting; oh we're getting a special deal." A lot of it because those people had to go legal, and they got brand new rates at 100%. Well those of us who had been in the industry for a while, typically were paying well above 100% because our competitors were not insured and they were part of the "Oregon Plan" so when they get hurt, the worker got insured and it was us that was paying that benefit to the Hoedad's and others. That's one of the reasons . . .

Brinda: You'll have to slow down here and back up and explain that.

Bob: Oh okay. In the Oregon Plan, there's no worker that's going to get hurt on a job.

Brinda: The Oregon Health Plan?

Bob: The Oregon compensation insurance.

Brinda: Oregon Comp. Oh okay.

Bob: It was called the Oregon Plan at the time. So if you had a bunch of Hoedads out there and they weren't paying compensation or had 12 Mexicans and they're paying compensation on three Jose's and whoever's hurt is "Jose." Those people, when they get hurt and they go to the hospital they're covered even though their employer's not doing it. The Hoedad's big shifty move was that they're all independent contractors, which was a crock. But it kept them from having to pay comp, so they go in there as an independent contractor that was injured on the job. And people like me had to pay a premium to cover uninsured workers.

Brinda: Okay. 'Cause just the rates just go up as there's . . .?

Bob: Yep. And then there's a high injury rate.

Brinda: Mm-hmm (affirmative).

Bob: The video I'll get you -- we did for them shows the different types of job descriptions and it shows people actually doing them. But the idea was we gotta show that a lot of these jobs are a lot safer than logging 'cause we were paying loggers' rates.

Brinda: Oh, wow. Wow. That's not --

Bob: So, and that was debilitating.

Brinda: Yeah.

Bob: You can go to Washington state, and it would just be a fraction. You could go to Idaho and it'd just be a fraction. Oregon was just really high.

Brinda: Did that change, the loggers . . . I mean were you always paying loggers rates?

Bob: Yeah. I was intervening on behalf of my business and ARC [Associated Reforestation Contractors] and that was it. We got variable rates.

Brinda: You did.

Bob: And I think they still have variable rates. It made it, for those of us who that were legit, it was making us more competitive. It lowered our overall premiums.

Brinda: Right.

Bob: The other thing we were trying to do, around the same time as unionizing, is expose the noncompliant contractors. That's what they were called.

Brinda: The noncompliant contractors?

Bob: Yep.

Brinda: And so that would include both co-ops as well as contractors who didn't pay workers comp?

Bob: Noncomplying. They were not following the law. They were circumventing the law. The Hoedads, they were being self-righteous about it, but it doesn't matter. It was unethical.

Brinda: Right.

Bob: And the Mexicans and the Russians were being clear as a bell. It was like okay; here's how you can compete; here's where you get another advantage.

4. [0:32:45] Industrial Reforestation Contracts

Brinda: Okay. So those work forces eventually moved on to the private land and displaced your work force?

Bob: Yes. Yes, 'cause those private lands had less and less need, and there's more and more surplus, and there was more and more . . .

Brinda: They had less and less need why?

Bob: Because of the housing market and that. They weren't logging. Then they started doing export and then they started like changing ownerships. The ownership of a Weyerhaeuser or a Publishers, is pretty well dedicated to, or Georgia Pacific, to getting that next profit from a plant . . . They want it, it's a good investment. Especially Publishers, they kind of sold themselves over much; they invested a lot . . . So a lot of crews became dependent, even Pihlstrom and the others stopped doing federal work and they started doing Publishers.

Brinda: Publishers is a company?

Bob: Yes, it puts out the *Los Angeles Times* . . . They came up to Oregon, and they bought a lot of small land ownerships in the 60s and 70s. Their big idea was to go through all these fields, and all these alder stands, and all these logged off lands, because there was such a high site. That means they grow timber real well.

Brinda: Okay.

Bob: And plant trees. And that way they would transfer wealth down forty or fifty years. They'd have plenty of fiber for their newspapers and they would take the surplus wealth from the 70s and 80s and send it on to the next century in growing timber. In fact, not only send out wealth but expand our wealth. But they were a little bit too optimistic. They didn't take into account animals. Animal damage to seedling numbers. And so their investments became greater, they were paying too much money for land, and ultimately it just didn't pencil out the way they . . . They could see after about ten years like, whoops!

Brinda: So they said that in terms of what, mortality of the seedlings or?

Bob: I think that was part of it. Maintenance of the seedlings, risk, the brush, animals, fire. But mostly I think it's the very real cost of long-term investments. Tying up a lot of capital for forty years. And who could perceive the iPad five years ago?

Brinda: Right.

Bob: So it's bad decision making. They kept assuming that there would be more and more demand for wood products and the value would keep escalating. And then the housing market of the 80s kind of like took those realities.

Brinda: So what's happened on those private lands now, in terms of forest management?

Bob: A lot of them were abandoned in the 80s and that. And have developed wonderful stands of second growth trees that are now merchantable. And I think that's going to be one of the interesting things that hopefully research points to, is most of those lands are doing fabulously. If you compare the money spent on those lands, compared to the money spent on restoration -- any you have your associate studying that -- hopefully you can put your data together in such a way that there can be a comparison.

Brinda: Right.

Bob: Because there's a real astronomic disparity that I'm going to predict between the two. I think you're going to find out that reforestation expenses, even absurd levels like Publishers' or the Forest Service, ultimately pencil out. Especially if you look at them as a cost of doing business. But the restoration stuff, I think you're going to find is pretty nebulous.

Brinda: Mm-hmm (affirmative).

Bob: A lot of that work seemed to be designed by graduate students. No offense! [Laughter]

Brinda: With no on-the-ground experience. No, that's fine.

Bob: Very nice in theory, but then there's this *Aplodontia rufa* ["mountain beaver" or "boomer"] that goes through and eats all the seedlings, and so much for plan A, then what?

Brinda: Right. Well okay, in terms of just new ideas about what's happening on private land that's now ready to be harvested? It's ready to be harvested, but is there a demand for it to be harvested?

Bob: Oddly enough, the demand has shifted because of some of these different market factors and policies we're discussing. Where the big diameter trees that used to get the premium dollar are lesser value, than small diameter trees that can be readily chipped or peeled or knocked into two by fours. The demand for small diameter logs is I think the best it's been in like a hundred years. More than a hundred years. It used to be a huge demand because of all the . . . They called them "tie mills" for the railroad ties, before they got creosote online. Huge sawmill industry for those small second growth trees. But now they're in demand more for construction materials. It's not railroad ties anymore.

Brinda: So would you say there's still a lot of cutting taking place on private lands in the Northwest?

Bob: Define "a lot." I mean, compared to the past times?

Brinda: I guess in terms of providing employment for others . . .

Bob: Way down, that's way down.

Brinda: And in terms of if those lands are going to be cut, does that mean it's going to be replanted? I'm assuming . . .

Bob: Well they have to be replanted by law.

Brinda: By law, right?

Bob: Right. Now the cuts . . . There's a lot less employment because there's been a lot of technical development in just the last 20 or 30 years. So now you've got all these machines that can go in and do the work of two or three guys.

Brinda: Like feller bunchers going . . .

Bob: Feller bunchers exactly. They couldn't operate on these hillsides before, so we had to all do that by man power. So now you've got one guy on a feller buncher machine, and you've got the manufacturer, and the maintenance, and the mechanic, and all those guys instead of four cutters and somebody running a skidder and somebody running a cat. But you still have to chop down those trees so then it'd be a two-part . . . You'd have five or six people doing what one or two do now. And then they're thinning a lot of these stands off, you have this big argument for thinning, so there's no need for tree planting there.

You can get figures real easily from the IFA nursery and from D.L. Phipps, Oregon Department of Forestry nursery. And those records are just outstanding. I haven't looked at them for a couple decades, but they will tell you what the trends are.

Brinda: Okay, on private lands as well?

Bob: Yes, specifically on private lands. IFA, they're not the only private nursery, but they've got a handle on the private nursery industry. And the state of Oregon is responsible for state lands and private lands.

Brinda: Okay.

5. [0:39:28] "Restoration" vs. Restoration

Bob: So the feds, they used to have their own nurseries, when they were doing stuff like that, mostly they shut them down. Mostly their logging and reforestation work is . . . "restoration."

Brinda: Okay so yeah, when you were saying that the restoration work, the cost of it . . . You said, "Restoration costs are pretty nebulous."

Bob: Well the work is nebulous.

Brinda: The work.

Bob: The cost is almost all . . .

Brinda: High.

Bob: Yeah, generated by tax revenues.

Brinda: Okay, right.

Bob: It's a money sink; whereas reforestation work, theoretically, should be a cost of doing business. So if you're going to log those trees off, you better plant something back for the next generation. Rather than, jeez you've got all this bare dirt, what do we want to do? Raise goats, plant trees, no. The decision is already made. You're going to log in Oregon, you're required to plant. I think that if Publishers had done their math that way, they might still be in business. "This is a great investment," they've talked themselves into it. It doesn't pencil out very well that way.

So if you're looking at restoration work, typically, what is it? Typically, decommissioned roads, putting structure in creeks, moving rocks around, hand pulling weeds . . . I'm sure it all feels good at the end of the day but it's like leaf raking. And it's busy work. It's like where is the economic benefit to that?

Brinda: Right.

Bob: I mean from a long-term resource perspective. I'm sure there's economic benefit, maybe to the community or the workers out there doing it. But the odd thing that happened, when this period started where they started laying off all the loggers and the sawmill workers and reforestation workers, they started getting a lot of federal money and they started hiring people like out of work commercial fishermen.

Brinda: So "they" would be the federal government?

Bob: Yeah. And they'd give them $7.00 an hour jobs doing what we could do way better, way cheaper. Not that cheap an hour. For four days a week, then they'd teach them data entry for one day a week. It's like job retraining.

Brinda: Mm-hmm (affirmative).

Bob: But it was crazy. It's like why are you putting loggers and reforestation workers out of work, so you got a training ground for . . . I'm gonna be mean again . . . graduate students and displaced fishermen. But that's what it was. It was typically a college professor, I can even name one of them who got lots and lots of money from the federal government. and she just set up these employee training things, and she'd never . . . Zero ability. She'd never had employees.

Brinda: Mm-hmm (affirmative).

Bob: I don't know if she'd ever had a budget. And here she is training people, theoretically, with lots of government money, how to be competitive contractors.

Brinda: In restoration?

Bob: They call it restoration instead of reforestation, but most of the legitimate jobs it came up with were logging or reforestation jobs that we've done either routinely or avoided routinely.

Brinda: Okay so what then, happened to your workforce? Because there was a huge number of workers in logging that lost their jobs, as well as reforestation, right?

Bob: Yep.

Brinda: So did they go into these retraining programs?

Bob: My crews . . . We could see the end coming for about six months roughly. And we tried some desperate measure cuts and everything. But then it was just abrupt, okay there's no more work. There was enough for five or six of us to try and keep paying debts and hopefully in a year or two this finally ends. We went and toughed out about three or four bad economic years, well, until it got ridiculous. And we just found some way to keep it going . . . So a lot of my workers joined other crews.

Brinda: What kind of crews?

Bob: Similar crews that were already cutting their wages and had different kinds of inefficiencies, like the three Jose's had going, or McCready's . . . There were some smaller crews that traveled around more, so that was their advantage. So doing the same types of work, but it was kind of like . . . I think a lot of people who worked for me had a big advantage because they'd been doing it for a lot of years, and I think they were really well trained. And so they were able to step into other crew situations and sparkle just from the beginning. So they'd get good wages, but good wages wasn't 14 an hour anymore, it was like 7.75 or something. So they'd take huge cuts in pay.

Brinda: Wow. So at that point . . . This is what, the mid-80s? Early to mid-90s?

Bob: Yeah, let's see. I think the mid to late 80s . . . That's where things got pretty desperate.

Brinda: And at that point, there was still work in planting?

Bob: Yeah, but it was way down. People had to work on Christmas tree farms to work out through the rest of the year. There was way less pre-commercial thinning. The planting pay was just a lot worse. Way less chemical application, the feds had cut it off entirely, and so on.

Brinda: And so now, the crews, what do you call these crews who are out there? Are they restoration crews, because you're right, they're not doing planting in the traditional sense.

Bob: I would say they're . . .

Brinda: They're doing a lot of thinning, they're doing the piling, the burning.

Bob: Yeah, hand piling. A lot of busy work. I'd think restoration crews would be a pretty reasonable thing for them. I think there are still reforestation crews around . . . But I think they . . . And there is where [John] Foster and [Tom] McCready would have a better handle on that than me. We reactivated Phoenix Reforestation, for example, but we're focusing almost entirely on a handful of people doing oak savanna restoration.

Brinda: So you're also doing restoration work now?

Bob: Yes.

Brinda: And is that on public lands or?

Bob: Private.

Brinda: Private lands?

Bob: Yeah. I am going to be putting out a bid for public land here in the next several days, though [crosstalk 00:45:45]. Well, I don't know. Sometimes you get what you want, that's the plan.

Brinda: Right, so what do you think of the work you're doing now? Do you see that as busy work?

Bob: Oh no. I'm adverse to busy work. If I see something as busy work, then I avoid it. There's a reason that I haven't necessarily done well economically through a lot of years. [Laughs]

Brinda: Clarify for me then, because before . . .

Bob: Busy work?

Brinda: . . . when, well not busy work but you said restoration now is shifting around . . . I mean, correct me if I'm wrong, but my understanding what you were saying is it's leaf raking, et cetera, et cetera.

Bob: Right, I've just got no interest in that. What we used to do a lot, more than other crews, they would go from conifer to conifer. So they'd log Douglas fir and then they'd plant Douglas fir.

Brinda: Correct.

Bob: And Douglas fir's main competitive around other trees, as it just kills everything. It just shades it all out. So you've got mushrooms, so if you log it real clean, there's really not much site prep. And it depends on, you can log a whole tree, whole tree type logging, or helicopter logging take the tops and limbs off. Or you can leave that stuff around, and you have to burn it, but the treatments are pretty straight forward. It's not like rocket science; it's like a pretty straight forward thing.

What we did was a "conversion" reforestation. We went from an 1868 wildfire that killed most the timber. Reprod was the second growth. Most of the second growth was logged off, came up into alder. The trees that were left behind were limby, there were also snags left behind, and we cut the alder as much as we could, we sent it in for chips and we sprayed it with herbicide, then we burned it, then we planted the Douglas fir. So we were dealing with a lot of sprouting trees and shrubs, a lot of understory, because the alder is deciduous. And so we were going from alder to Douglas fir.

And we got a lot of heat in the press from the environmentalists because they were saying, oh look we're creating tree farms. And we were actually going back to the original conifer pattern that had been there 400 years ago.

Brinda: And this is ultimately for commercial use?

Bob: Yes. It was large land owners, Georgia Pacific, Publishers, Weyerhaeuser. That was kind of what we focused on because we could do it all. We could cut, spray, fire trail, burn, plant, replant, animal trap. Anything that was needed, we had trained crews of full-time professionals that could go in and do the job it expects. And no problems. Schedule the work, and we'd been doing it for decades.

Brinda: Okay, with the ultimate goal of having a resource to harvest and start the cycle again.

Bob: Right, yeah. We were growing timber. Now we are going to areas that are covered with timber, we're cutting it away and we're saying, it's the oak savanna, the forbs and the bluebird habitat, and the endangered butterfly habitat. That's endangered. We've got plenty of timber out there. Let's get these firs out of the way. Now they are the weeds. And let's restore oak savanna type habitats. So that's kind of what we're working on now. Let's restore traditional type burning practices that the Indians did two and three hundred years ago, to meet modern management objectives.

Brinda: Okay so basically, you sort of meet a change in objective.

Bob: Yes.

Brinda: Okay. And how do you feel about that?

Bob: I think I'm underpaid, like everybody. Not enough time off, not enough benefits, not enough pay.

Brinda: But the work is on private land. And the owners are saying, to oak savanna?

Bob: Yeah.

Brinda: They are saying that, are they? I'm just a little confused.

Bob: Okay, the private land owners, there's only about one percent of land left in that condition that can be restored. Almost all of it's on private land. So if the private people don't do it, then nobody can. But I'm also going with proposals to the Forest Service and the BLM, where they've had these massive, catastrophic fires. My PhD is in catastrophic fire, and my training is in reforestation. And so I'm putting out different kinds of prescriptions that acknowledge the need for a more diverse wildlife habitat in the landscape but also insists on an understanding that seems to have been entirely missed. These are management problems. They need people out there doing the work to create these conditions. It's not a benign neglect "non-declining, even flow, naturally functioning ecosystem."

And that's what they've been arguing: first, on the spotted owl, and now they're arguing that on oak savanna. That they can kind of restore "natural functions," then they get a little vague real quick. And by doing that, that then the thing will continue with recycling carbon and then there'll be nutrients in a "non-declining, even flow, naturally functioning" fashion.

Brinda: Where does the funding come from?

Bob: The funding to do that kind of stuff?

Brinda: Right. For your . . .

Bob: Taxes . . . Oh, for mine? Mine's private. So it comes from . . . 40 acres I just designed, 22 acres is designated for one acre home sides, residential home sides, because they can hold habitats sufficient for species with limited range. And 18 acres is designed as a hundred-year outdoor school room for local students. So how do you get 40 acres into a way that pays for itself and can be managed through time? So those are the questions I've got. That's what keeps it interesting.

Brinda: And the private land owner too, are they developing their land?

Bob: In this particular project, that's the idea. They'll develop 22 acres as residential, using native plants, and native animals, and 18 acres will be developed as an oak savanna restoration project; go back and create an earlier tradition.

Brinda: But that would be part of a communal land [crosstalk 00:51:58].

Bob: Part of the city of Corvallis . . . Yep. Open space in the city of Corvallis. So I'm planning an urban, privately owned property, with residential development with a long-term study, and not for recreation.

Brinda: And so is that becoming a public easement?

Bob: Well that would probably be most likely. Some kind of conservation easement that would be going to a non-profit. Something of that nature.

Brinda: Mm-hmm (affirmative). So what's happening, would you say a trend that's happening is that some smaller private land owners who perhaps had timber land, are now developing that. And so rather than planting for timber, it's okay, let's do the restoration thing but you also have development because it's more profitable for them?

Bob: Yeah. A lot of them are buying into the idea, and there's a lot of federal programs right now which allows them to do it. They'll get federal cost share to create all of that habitat. But there's no follow up. There's no monitoring, there's no maintenance. And so my fear is it will be like, that's what I'm calling leaf raking. It's like, you can do all that work, and then if you don't maintain it, the leaves fall down and look what's happened to them. Leaf raking.

Brinda: Right, I mean you studied that, like a lot of these oak savannas consistently manage to be oak savanna with the right practices.

Bob: Yeah. And so that's what we're trying to get going. We're trying to get burning back in the community; we're going to have difficulties doing that, so it's educational, it's instructive, but we're also on the cutting edge. We're not reflecting a market trend. We're reflecting a potential management direction and we're highlighting the costs and benefits of that direction. And working with school kids to do that work. So I'm trying to use my degree as an educator.

6. [0:53:54] "Conservation" vs. Conservation & Jabberwocky

Brinda: Okay, sure. And just your reflections on then, going from planting and sort of a fairly vibrant timber-based economy, and resource-based economy to something that's let's say, more conservation oriented, restoration of the landscape. Economically. Is the restoration economy a sustainable economy?

Bob: If we don't call it conservation, it isn't.

Brinda: If you don't call it . . . Okay.

Bob: They're starting to call themselves conservationists.

Brinda: Who is they?

Bob: The preservation community, which, my attitude has shifted on who those people are and why are they doing these things. But when they start calling themselves conservationists, they are misrepresenting themselves. And misrepresenting themselves is like Hoedads claiming themselves as "quality control." It's like, other than the fact it's not true, it sounds good. So it's a promotional thing that they've been using consciously. Conservationists go back to Pinchot a hundred years ago, and it's "wise use." It's like, first and foremost, the local industries. Well these people aren't reflecting that philosophy by any shape. Theirs is an anti-logging philosophy fairly consistently.

Brinda: Okay.

Bob: So the idea is to not log riparian zones, let the conifers somehow create salmon, not log the burned-off lands, have the snags be wildlife habitat and develop soil structure. And so to marginalize or even eliminate the timber industry. I think that's pretty fair. However, we're all living in wood houses, we're using wood bookcases, books are wood products, so is newspaper, toilet paper, paper bags. We're using wood furniture, so there's an hypocrisy involved there. Where do we get our wood products? Well then, now we get them from Canada, which Weyerhaeuser also owns. So there's an hypocrisy there in Weyerhaeuser fighting to keep the environmental laws going here because they make good money in Canada, and they get to say they eliminate the federal government as a competitor.

So the artificial elimination of wood products over half the land in Oregon, and I think I can say the same thing for Washington, has been a little bit muted because there's been all this upswing in technology. Hewlett-Packard and these other outlets, Intel, that moved into Oregon, they're the largest employers, and then agriculture. Hewlett-Packard in Corvallis and Intel in Washington county, I think they're the largest employer in the state.

Brinda: Mm-hmm (affirmative).

Bob: And they're just California high-tech spinoffs and now they're our major industry. So I think it's been debilitating to the school system, to the rural economies, and it's not sustainable.

Brinda: What is not, the restoration?

Bob: The restoration movement, as I've defined it. [Laughs]

Brinda: Okay, and yet you recognize that you're a part of it. I mean just in terms of the work you're doing is restoration though.

Bob: Yeah, I'm doing restoration work and that makes me an anomaly in the restoration industry. I don't think most of them are doing restoration work.

Brinda: Okay can you say more about that? That's why I'm confused because . . .

Bob: Right, I think most restoration work is not going in and restoring an earlier condition. It's creating an unprecedented condition. So it's not really restoration. They call it restoration, just like they call it conservation, but they're really not conserving resources. They're not using them at all. I would say they are wasting resources but they're saying they've got alternative values, like wildlife habitat or aesthetics or soil or something of that nature. So the restorationists have backed away from the position that they are restoring kind of sets of conditions. Because I'm an historian, and they're not. They're just making up . . .

Brinda: So what are they doing? What are you saying they . . .

Bob: Now their rationales are "restoring functionality to an ecosystem."

Brinda: Okay.

Bob: Are you familiar with the poem Jabberwocky?

Brinda: Mm-hmm (affirmative). Yeah, it's been a while.

Bob: Okay, but Lewis Carroll is basically using nonsense words to sound as if he's actually making a speech.

Brinda: Okay.

Bob: When they say they are restoring environments for the restoration movement . . .

Brinda: And who is part of this movement, when we're talking about "they?"

Bob: Well, you've got to be more specific. I would say it's fueled by very successful lawyers because they're dealing with federal policies.

Brinda: And we're talking about public land, on federal land?

Bob: Yeah, and that's over half the land in Oregon.

Brinda: Right.

Bob: So it's like, that's why I have to stop on some of these questions, I mean realizing, well I'm talking just about the federal land, or we're talking just about the private land. There is very little restoration going on on private lands.

Brinda: Okay.

Bob: And what's being done on private lands is typically paid for with government money. Fish and Wildlife is paying for habitat. Or Soil Conservation Service is paying for streambank stability. So they're getting federal money, they are getting cost-share grants to plant seedlings in riparian areas for shade.

Brinda: On private lands.

Bob: And on federal lands.

Brinda: So that money from the federal government, or these agencies in the federal government, help subsidize your work on private land.

Bob: I don't do that work though.

Brinda: Oh wow, okay.

Bob: See that's what I'm saying is like, that's the busy work.

Brinda: I thought that that's what you were doing on private land?

Bob: No, no. We're using the same word. When I'm saying restoring, I'm talking about restoring an historical condition.

Brinda: Correct.

Bob: Or precontact condition. A combination of plants, animals, and management processes that accurately, as much as possible, reflect past conditions.

Brinda: And is part of that then paid for . . .

Bob: It's all been private money, so far.

Brinda: It's all been private money? They're not taking from . . .

Bob: I'm not using tax money to do this.

Brinda: Okay. You might not be but the private land owners might be.

Bob: Well but see, you're asking me what I'm doing for restoration so I'm . . . I would probably accept federal money, and we're looking to do the same types of work on federal lands, but most of the land is . . . okay, is like the Oregon Watershed Enhancement Board to the state. And then that money is lottery revenues or it's from the feds. Or it's from Fish and Wildlife Service. So there's a lot of federal money going into private lands to achieve ecosystem "enhancement," which means like the whole area has given objectives. And ecosystem objectives are typically open culverts for the fish, plantings along stream banks, movement of root wads to create pools, and this is what's called restoration activities. Or they're being performed on federal lands, and road decommissioning is called restoration. But also putting root wads in creeks and moving rocks around and that type of thing.

They just had a front page story where they took a bunch of logs, that you could build a house from, but they jammed them in Calapooia River, where I live, and the person said, "Okay, in the next flood this is gonna shift around, it's gonna create pools for the fish." Which is probably true. But they could've taken a backhoe and created pools for the fish with about one tenth of the dollars. At some point what's going to happen to the logs, is they're going to blow out in a flood. That's what I do with my research is how often do these pulses occur then blow out all the logs and they're gonna be somewhere around the ocean or downtown and they're gonna cause a lotta damage like a big Brillo pad as they're going up and down. That's the part that the restorationists, "they" that I'm talking about, don't get, and that would be the policy people, the legal people that are fighting for not having logging, having restoration, brush piling, thinning, what they call thinning, instead.

So we've got the policy makers, we've got the lawyers that are fighting for it, using the federal laws and regulations, sometimes even odd rural employment type acts. Then you've got the movement people themselves. These are the people that climb the trees or work the jobs. I think you'll find a fairly consistent philosophy, either they're labor-oriented and they're sending out Mexican work crews trying to make the dollar for whatever you tell 'em to do, doesn't matter, paint the porch, move the rock, thin the trees, pile the brush, casual labor crews just do whatever they're told -- or they're idealistic in that they're people that think they're out there helping the fish and helping the environment.

Brinda: Right, and those are probably two different groups 'cause what --

Bob: But they're both, I think, reflecting of current restoration activities, and they'd both be dependent on tax money.

<u>7. [1:03:31] The Environmentalists & Precontact Conditions</u>

Brinda: Okay, and so what a lot of I've been hearing just in terms of reading from a lot of these same movement people, and here I'm not counting the underground crews 'cause what I've found, at least through interviews, is say the Hispanic workers or the contractors, they're not necessarily advocating for any type of work, they'll just do what's there.

Bob: Right, that's why I'm calling them casual labor work crews. Exactly.

Brinda: Okay, versus maybe more movement people who --

Bob: They're better educated for the most part, I think.

Brinda: The movement people or the --

Bob: Yeah.

Brinda: Yes. They're sort of maybe having a more sort of idealistic or ideological agenda of what they want to see and based on the policy changes, crews are doing whatever work is there.

Bob: Right.

Brinda: Okay, but what I'm reading, and the papers say is apart from the we're creating "ecosystem functionality," we want the salmon back, I often see a justification as also, we are creating sort of pre-settlement conditions and I'm hearing you say that.

Bob: I am saying . . . we are . . . I am doing it, but I have not seen a restorationist do that. I have seen them claim it.

Brinda: But have you heard them say that?

Bob: Yeah, I've heard 'em say they're conservationists too. That's why I'm saying what we're doing is different from them, as they're claiming.

Brinda: 'Cause you probably have things that are restoring ecosystem functions and ecological . . . I mean, what's the difference, I guess?

Bob: One of the differences is, okay, when we say we're restoring precontact conditions, those are very definite conditions, using research skills to determine what those were. That's what I do.

Brinda: Right.

Bob: Okay, when they say "restoring ecosystem function," that's Jabberwocky. What the hell does that mean? It's like Jerry Franklin was saying, "We can't log any of these trees off 'cause this is all the wood the soil's ever going to get." Well, maybe in his lifetime, but he's old. But what does that mean? Does soil need, does it need aspirin, too? Why do these coho need conifers? They didn't use to. Why do they need them now? Do they need running pools? Okay, but what's wrong with just making a pool for 'em if that's what they need?

Why make a policy for them that's really inefficient under the assumption that's what they need without at least some scientific testing. So there's not scientific management going on. It's a feel-good, leaf-raking, ephemeral process that I'm referring to that they get either the labor crews doing, and making a profit for themselves doing it, and they could care less what it is; or we get the idealistic people promoting, supporting, protesting, hiring lawyers, filing suits, so there's a real litigation aspect to forest management that is a large share of why reforestation industry has changed so much.

Brinda: Okay, then in terms of precontact conditions, a lot of environmentalists do use those --

Bob: Words.

Brinda: -- words.

Bob: Yep.

Brinda: And what are some of the justifications for precontact conditions?

Bob: I just wrote a nice little article on this. I should send it to you. It's intended for a general audience. The main difference, and this causes consternation with people when I bring it up, but I think it's exactly right: what they're talking about in precontact conditions is a basic form of institutionalized racism. They are describing the environmental movement and that's been a whole lot of my antagonism towards them the last 10 or 15 years because they've been informed. They know better now.

Their idea is that the people for 10,000 years lived in this environment that were totally at the mercy of nature. They have an anthology written in 1999, that's only 6 years ago, by leading scientists all around the Pacific Northwest, including some at University of Oregon, stating that you cannot differentiate between Indian fire, which is my specialty, and lightning fire. Well, you can, but they couldn't, but they're the types of people, I mentioned Jerry Franklin or Jim Agee, or people like that that have got these ideas that Indian people were that incompetent, just waiting for lightning to clear some huckleberries for 'em or keep the savanna open, and that couldn't happen in the Coast Range where I studied 'cause there is no lightning fire.

It's one of the least incidents of lightning fire in the world, but it couldn't happen anyhow. All the oak would disappear, the camas would disappear. Tens of thousands of contiguous acres of tarweed would disappear. They'd all get overtaken by conifers if you didn't have people in the environment like we did, disturbing it and my specialty, burning it. But they're also tilling, pruning, and walking and [unintelligible]. What we call structure and creeks, they called firewood or impediments to travel.

People didn't let stuff build up like that. They had canoes running up and down the streams, and they had fire going constantly. They didn't let wood float by and build up on the beach as salamander habitat or any of that, clam colonies, or whatever the heck they're talking about. The area was pretty much bereft of dead wood where people lived, especially where they lived for thousands of years.

When you've got weirs that are 3,000 years old and village sites that are 8,000 and 10,000 years old, so those areas people use wood every day. It couldn't have built up the way that they're saying, so I think that there's this idea that a functionality outside of human imposition of management . . .

Then the other side of the coin is like then everything we've done, "we've" been mostly whites for the last 150 years, has been evil. [Laughs] We've logged, we've plowed, we've fenced, we've put in pavement. Well, that's all true. I don't know why that's evil. It like everybody seems to be running their bikes down the pavement and buying their groceries at Safeway and living in a nice house, so it's like how that constitutes evil; it gets into why I think a whole lot of the movement's based on which is more political agenda than science.

I think that would be a way to differentiate between reforestation workers, or basically trained professionals, using scientific management processes for resource management; using scientific methodology for resource management as opposed to a lot of the restoration movement, which is based on almost like spiritual guidance. They're talking about precontact conditions, but it's like this idealized environment in which people are inconsequential and to me that's not idealized, that's not only erroneous, but it's also racist. They don't seem to have much understanding or respect for what people were actually doing here for last 10,000 years, which is --

Brinda: Okay, I think I'm starting to understand the distinction. You're saying some of the way the environmentalists use this word or concept of precontact conditions is like writing the people out of that environment, writing the managers, the active management of these lands out.

Bob: Yeah. Right. They'll put the people in. They'll pay 'em lip service but they don't have them as a relevant factor. And they're judged by "naturally functioning ecosystems." That's why I keep pausing here. It's like where did we get off into nature versus nurture? I think it's naturally functioning, but I think it's only natural that somebody's gonna want to drive a car to Roseburg instead of walk, especially in this rain.

So I think those are natural things and I see humans as natural parts of the environment and I see us as particularly inventive and industrious; but bears use tools, ants and termites build homes, so do birds, they build nests. We're not the only ones to use tools or feather our own nests. We're just way better at it than any other animals so far.

You take people out of the environment, and then you start looking at that as an ideal, and to me that's just bad policy. It's 'cause they're not gonna achieve what they want and if they did, why don't they just move to Mercury or Mars, so it'll be without people where there are ecosystems that don't have human involvement. So they're idealizing to me the wrong set of conditions and it's a bad religion for that reason.

Brinda: What would you think would be an economically viable set of conditions? What would your vision of that be?

Bob: I think Indian-type burning conditions, if they were re-implemented across the environment, where you have seasonal burning, instead of putting out fires in August and September, you're out starting fires, but they're in areas in which you've set the fuel loads, in which you can maintain X amount of habitat, and your justification for that might be recreation, hunting, photography, something like that because -- maybe even native foods.

Maybe there is a market for acorns or camas. There's a developing market for camas or cat's ears or other wild strawberries that are native, or wild blackberries that are native. There might be developing markets for those foods as there had been for 10,000 years, maybe at pow wows or, I don't know, traditional restaurants or, I don't know. That's possible, but in the interim all we're getting is a lot of diversity and then we've got to keep managing it. That creates jobs and --

Brinda: Diversity of terms of . . .

Bob: Plants and animals. If you have a Douglas fir environment, which grows your most timber and creates your most money, within about 10 or 15 years it shades everything else underneath, it drops a bunch of needles on everything, it becomes very acidic and you've got some mushrooms; but you don't hear songbirds, you don't see deer, you don't see people. It's dark and dank and it stays that way for a hundred years and then it starts breaking apart and you have bits and pieces, but you also have falling trees and limbs and those things, so it's a dangerous, shady, cumbersome environment as it gets older.

But oak, you have a lotta daylight. You might have 30 species of forbs associated with oak woodlands. You might have half a dozen different type of bulbs. You might have half a dozen native grasses. You have, I don't know, 100-150 different types of birds, not three. So by diversity, in an oak woodland, you have significant increase in vascular plant species and vertebrates, but also bugs, butterflies, things like that.

That's what I mean, it's like it's almost a park-like condition. It's highly productive for crops that are no longer commercial, but the environment is productive for humans values, and so they serve as firebreaks. There's ways to, I think, enhance timber production by isolating it into the areas where timber had been grown 150 years ago. I think one reason it was not all burned up was because these other areas were burnt regularly and there wasn't a chance for ladder fuels to build up and there were buffers between them, and when an area would burst into flames, you'd get a crown fire, the animals would have these meadows and prairies and bracken fern strips and things to run to and we know that's what they did. I remember records of that in my own family from 1902 that people would just go into these Indian prairies, people and wild animals, and wait for the fire to burn through.

Brinda: It sounds like there is, well I don't know if there's a need right now. It sounds like the timber industry, at least in this area, has gone downhill pretty quickly.

Bob: Yep.

Brinda: But you're advocating sort of both types of resources.

Bob: Yep. More of a balance but a type of pattern that was here 150 years ago, restoring older landscape-scale patterns I think would meet a lot of the current objectives and would also create a sustainable economic situation.

Brinda: And where do you think funding from that might come from?

Bob: Timber. One of the nice things when you log trees is they produce income, but then the other ones are just for talking about. Maybe hunting for meds, maybe recreational camping fees, maybe incidental harvests, specialty crop harvests like they do with mushrooms now, maybe camas, bracken fern, things like that. So probably incidental use fees, incidental secondary forest products, but principally timber production pays the bills, just like football pays the bills at the college.

Brinda: Yeah, but timber production might pay the bills, might pay for this restoration on other areas, sort of non-timber-producing areas.

Bob: Right.

Brinda: Okay.

Bob: And the rationale would be they're trying to grow, I don't know, spotted owls and strawberries and everything in the same acres. They want X number of snags, X number of conifers per running stream mile and so on. If different areas were relegated to different things so that the endangered species were over here with oak and loggers are over here with the Douglas fir and the tree planters are burning out the woodlands on one day and replanting the Douglas fir on another day, then I think we'd get that balance of values and we'd get the efficiency of a trained workforce and we'd reduce the dangers of wildfire, increase the endangered species habitat, timber production, and so on. It's based on active management as opposed to benign neglect which is what wilderness areas in --

Brinda: So, yeah. What I'm hearing you say, and I think I'm understanding more of, is that some kinds of restoration work, what they're calling . . . Let's say "they" as some environmentalists --

Bob: The environmental industry. [Laughs]

Brinda: Yeah, the environmental industry, and again that in itself seems really complex to me 'cause I hear different voices even in that industry. It's hard to characterize it as one, but folks saying restoration work is almost like leave it alone, don't touch it. Where some people, environmentalists, are also saying no, it has to be actively managed and burned but according to sort of Indian burning practices, but who might say, something like you, still need to put logs in the streams and all. So it seems like there's a whole variety of management practices some of which overlap, some of which don't and it's hard to categorize.

Bob: I think the second group you're talking about is an emerging group.

Brinda: Is an emerging group? Okay.

Bob: It's the politically correct environmentalists who are saying, okay, we were wrong about the Indians, we need 'em there, but we need to keep 'em in their place, around the heads of creeks and here and there, but we still need to have this old organic wood material, we still need this . . . and then they go right back to a "naturally functioning ecosystem" with, what did they used to call 'em? Not figureheads but . . .

Brinda: Tokens?

Bob: Tokens, yep. Token people in the environment. Token village here, token campground there. I've had anthropologists say, "Yeah, but they didn't go up in the hills." It's like, "Well we do, why didn't they?" It's like, "How could they carry water up there?" It's like, "Well, you go down to the creek, you drink, and then you walk up to the top of the hill." It's like we did it all the time tree planting. You didn't carry a canteen around. You just carried it in your stomach like anybody.

It's like, "Well, why would they go up there?" It's like, "Well, nice place to meet people, beautiful view, something to do in the afternoon, and just happens to have all kinds of food crops up there." So we know not only why they went there, we know what time of year they went there, when you harvest these crops, when you pick these berries, when you harvest these acorns.

We start seeing that people don't have seasonal rounds, they're just like farmers that are tending crops, and that's the pattern I'm finding a lot in western Oregon, not so much eastern Oregon. And well, there's so much reforestation work back there 'cause now it's like junipers encroaching into prairies and people there really probably did have seasonal rounds and bitter winters and things. But west of the Cascades, where the principal timber grows in the world really, say from Shasta right up into BC, the interior valleys, from Sacramento Valley right down to Mexico, really had pretty large populations of relatively sedentary people that practiced more farming than forestry, more farming than hunting or gathering.

They weren't really hunters and gatherers, that's characterizing them badly. So, anyhow, that's a long way to say that even though they're now saying lip service to these Indians, I'm not familiar with any . . . In my field of study I'm not seeing any literature coming from them. I'm not seeing anything that's particularly revealing that they understand the topic. My field of ecology is people in the environment. When they're putting people in the environment it's usually really awkwardly 'cause they haven't done it before and so they're trying to treat people; they go okay, these people were spiritually advanced, and they had oral traditions, and so they've got this mythical human being that they can kind of put in the environment, that can kind of subsist, and I think that's just not an accurate viewpoint. I don't think it's very respectful either.

Brinda: Yeah, it's almost like this idealized or romanticized vision of --

Bob: -- of a eunuch without a brain; [Laughs] some simple little human that how did they get by, dang it? They're so soft and gentle on the earth.

Brinda: Right, the noble savage.

Bob: Yep. Their sole viewpoint, the gentle nitwit. It's like it's not a very noble being that they're describing, they're describing somebody that's virtually incompetent and that's why I think the institutionalized racism comes in. It's like, "Hmm, too bad they're not competent like us, they could've got a job logging or something."

Brinda: Huh. I don't know if this is part of your thesis. You were looking at management practices, right? Pre-settlement?

Bob: Mm-hmm (affirmative).

Brinda: And what the land used to look like?

Bob: Yes, as a result of those management practices.

Brinda: As a result, and so I've heard about or read about Muir taking his horse through park-like conditions, Ponderosa pine, and it was actually not very dense, like newer.

Bob: Yep, right.

Brinda: So, is that what you found as well?

Bob: Mm-hmm (affirmative). Virtually no snags, no coarse woody debris, no structure in the streams. They were like they had--

Brinda: In certain parts of Oregon, right?

Bob: Almost all of it.

Brinda: Okay.

Bob: There were large areas that were forested, but they're almost topographically isolated, broken away from . . . like the east slope of the Coast Range was, the whole entire east slope was virtually unforested, so they started calling it Willamette Valley, like it wasn't a range of mountains. Well, it was forested with oak and just little patches of fir. The west slope was virtually all forested except all the ridges had bracken fern prairies, all the peaks were bald, and all the lowland valleys and low gradient areas where the fish were, were grass and prairies.

So, yeah. The conditions Muir described were conditions created by constant incremental and cumulative gathering of firewood. That's a massive amount of coarse woody debris everywhere that grandma could send you to get it. Every day, you cooked lunch so every day firewood was gathered and there is the burnout around towns and campgrounds and homes, so you had to go outside of that area 'cause if you didn't gather the firewood you burnt it off so you didn't want to waste it, so you gathered it. Anything came floating down the stream, that's just like today.

Brinda: And then sort of the buildup has taken place because over the years there was just that fear of fire and --

Bob: Well, the Indians died off very dramatically in the 1780s, and then by the 1830s --

Brinda: Through disease and warfare.

Bob: Not warfare, disease. In northern California and southwest Oregon, there's quite a few tribes that lived there. They were isolated and they were killed off by white people in the '50s, 1850s. But, by the 1830s, most of the people of western Oregon had died from disease, smallpox, malaria, especially.

Brinda: Brought over from the --

Bob: Yep, Africa and Europe, and from the United States, which is the east coast. You had white and black people coming here that are bringing diseases in. I think the Chinese didn't start coming till the 1860s, but by that time most of the Indians were gone. So in western Oregon, not so much up in western Washington, and not so much down in northwest California, but in western Oregon most everybody was dead.

8. [1:25:53] Fire Suppression vs. Fire Exclusion

Bob: So, when [white] people came in the environment, it only takes 5 or 10 years for the brush to build up, and what I've been looking at the Biscuit Fire here, about 30 or 40 years for the ladder fuels to build up, and about 40 to 80 years for Douglas fir to overtake oak. So you've got basically somewhere between 10 years and 80 years to create this transition to a firebomb, where you've got ladder fuels, and you've got contiguous fuels along the stream banks.

Brinda: Well and there was like a massive program of forest fire suppression as well.

Bob: Yeah, but that's way overblown. The bigger problem was fire exclusion. They stopped people from burning. They took fire out of the woods, whereas fire suppression says, okay, you've got a fire, now suppress it. You can't suppress a fire. Every one of these, why are we having 500,000 acre fires a hundred years later? We've got satellites and we've got air drops, and we've got Caterpillars, and we've got hoses, and we've got internet communications, and we still can't put a fire out. The weather has gotta change. So the fire suppression policy --

Brinda: Fire exclusion. Okay, a policy of fire exclusion.

Bob: Yep. First, get rid of the Indians, cause you gotta have livestock, so you want to graze. And then you get rid of livestock and people that are starting the fires, they called them fern burners in the 1930s, they would burn off these areas that the Indians had burnt off, and they stopped it. The Keep Oregon Green movement was a reaction to people setting fires so there's a big difference; 'cause with one of them you say okay, we want to return it the way it is, let's have a "let-it-burn" policy.

That's the assumption that people are nitwits, and fire would start and then everybody would go running around till it went out or something; rather than set fire to it, a prescribed fire policy, which is wait a minute, let's set fire to this now while we can control the smoke so there's not so much pollution and so we've created a fuel-free zone so if the wildfire comes from those mountain over there it's gonna hit like bare dirt.

Brinda: Right.

Bob: That type of purposeful fire exclusion is the big change that allowed all the ladder fuels and the shrubs to build up, I think.

Brinda: Right, but there is this policy of fire exclusion on plantation lands, right?

Bob: Well, there's a policy of fire exclusion on virtually all lands except campgrounds with firepits. There's some prescribed burning that's going on, but not even 30 years ago we did 18,000 acres of prescribed burning.

Brinda: Where?

Bob: On slash areas, areas that we had slashed and sprayed over on the coast and mostly in Lincoln County.

Brinda: And then after that you would plant?

Bob: Plant.

Brinda: Okay.

Bob: So we really learned the value of a good hot burn, and if you expose the mineral soil that all the environmentalists used to say was terrible, that's actually the very best you can do 'cause that's just what the Indians did. Once you got bare soil you got trees that'll take. They don't have competition, and you've burnt off the weeds and so on.

Brinda: Right. And I guess here the difference would be the Indians burned for basket weaving material or . . .

Bob: That'd be one.

Brinda: Okay.

Bob: Mostly food production.

Brinda: And food production.

Bob: They would keep the oak open cause gathering oak acorns was a major thing and then usually there's one or two crops in conjunction with it. In the low valley areas there's camas a lot and they're digging that up so not only are they burning it off, but they got bare dirt when they're through digging.

Brinda: What's camas?

Bob: Camas is a beautiful blue flower. It stands about that tall and it blooms in May, and it's got a bulb. It was like the equivalent, it would be the Irish potato, or, in your country, it's probably rice, or has it turned to wheat?

Brinda: No, the north is wheat, the south is rice.

Bob: Okay, so it'd be like wheat or rice. Now there was something called Indian wheat which was tarweed, which they roasted, and they had the seeds, but camas was a primary food. It would be like the staple for most of the northwest tribes. They try to tell us it's salmon, but it was camas. In some areas they didn't get salmon hardly at all. There were other roots and bulbs and things that people ate, bracken fern roots for starch.

Brinda: Okay, and they used wood primarily for firewood? Is that accurate?

Bob: I'd say yeah. I would say yeah. They also used large wood for . . . they had canoes that went out in the ocean. They could haul 25 and 30 people and they could go out when European ships couldn't, so they were able to travel on the ocean on large boats. They had houses that were big, 300 feet long, as long as a football field and wide, and maybe held 20 or 30 families in compartments with cedar plank siding to 'em, so they had a lot of use for large wood products. Then smaller wood products would be arrows and bows and spoons and bowls and lots of fibers, lots of woven materials.

Brinda: Okay, so in terms of life sort of coming back to the present, the present economy, do you see any move towards a restoration movement in terms; the restoration defined as . . .

Bob: As I'm defining it?

Brinda: As you're defining it.

Bob: I hope so. If they're going to keep their credibility, which they've been very credible with the courts, but they're losing credibility finally, quickly, with scientists.

Brinda: Really?

Bob: Yep. And I'm one of those examples. It's like I'm a scientist, so it's like, nope, these people are on a mission, they're not being consistent with their logic or their data. So they want one thing but their method of getting there is wrong. It's like they won't get there the way they're saying. So I'm hopeful that if the restoration movement continues, that it actually goes towards an historic range of variation and actually looks to restore actual conditions, and we forget about this functionality silliness 'cause they can't tell me what that means. You can't tell me; I can't tell you. It's like functionality is kinda like serious [unintelligible]. It's like well, what's that mean exactly?

Brinda: Okay.

Bob: And the way they're using it; that's kind of the Jabberwocky thing is . . . it's actually probably more like an *Animal Farm* thing where they say, "Oh, now we're conservationists" 'cause it's good to be conservationists, but it's like everything they're saying and doing isn't what conservationists say and do. Why are they calling themselves that? It's very convenient.

Now when they say they're restoring precontact conditions, that's not what the Indians did. Why are they saying that they're doing that? They're trying to sell us the idea that they know what those conditions are.

9. [1:33:15] Wildfire Risk & US Forest Service Accountability (2005)

Brinda: Right. Where do you see the Forest Service moving? What direction do you see the Forest Service moving in, in the management of public lands, public forests?

Bob: I hope they stop moving downhill really fast like they've been moving in the last 15 or 20 years. They're just going . . . There's a book that just came out from the old Forest Service guys and the problem with the Forest Service is, it was like a hockey team. It was all old white guys. So, they said okay, now we need diversity, so they just stopped hiring old white guys. So they got rid of their knowledge base, and they started laying them off through attrition, and they started replacing them with a sexually and racially diverse work crew, which is good. But, they did it so quick that the people didn't have a chance to learn their jobs. So they got a lot of political appointments basically. Not friends of the President, but friends of the Ecologists, maybe. They put together a well-educated, almost totally inexperienced work force. So people I think, like in the [*Evergreen*] magazine, I can show you where it's saying, "Hey, if we keep managing our forests like this they're all going to burst into flames." But that was a pretty good prediction.

And I also say, "but people won't stand for that." That's where I was wrong. People have stood for it, because the Forest Service has been incapable of harvesting those lands or regenerating them. They really kept a lot of people in the position of very little timber sales experience, or reforestation experience, or fire management experience, that are in there managing fires and making plans and planning for reforestation. So it's discouraging. So I don't know where they're going. I would like to see the Wilderness areas go over to Parks, because Parks are at least managing for a period of time. They're not just letting them all burn up, like the Forest Service is doing. I would prefer that the forest in Oregon were privatized over what the Forest Service has been doing. I think the Forest Service did pretty good for 80 or 100 years, but I think the last 10 years, they're just fighting litigation and that seems to be . . . I was going to say, the old guys, they just finished a book. It's a --

Brinda: Yeah, what is that?

Bob: "We have an objective." *We Had An Objective [In Mind: The US Forest Service in the Pacific Northwest, 1905 to 2005],* and--

Brinda: Who wrote that? Do you know?

Bob: Oh, the Old Smokeys. It was just a bunch of stories from the Forest Service guys, but I've talked to a couple of them and we kind of kid around. It was like, when you read that, the way they pronounce it now is, "we had an objective," but the way these guys titled it was, "we had an objective." [Laughter] In other words, like, we knew what we were doing.

Brinda: Right.

Bob: As opposed to now, so.

Brinda: Right.

Bob: It was kind of like an inside joke. [Laughs]

Brinda: And do you think BLM manages the land . . .

Bob: Much better, but their management constraints are different. And the main problem at BLM is that they've got this checkerboard pattern that they inherited from policy, and I think that if they had done something like they were threatening to do about ten years ago, like a land exchange where they could get more --

Brinda: Continuous?

Bob: -- continuous ownerships.

Brinda: Right.

Bob: That the need for roads passing all these areas and then the straight lines in the environment which are [unintelligible] and which are fire problems and so on. Those things would . . . could come to an end. I think that they've had a bad management situation imposed on them, but the management of their lands has been a lot more professional, and they've been managing them much better than --

Brinda: And when you --

Bob: -- fighting litigation.

Brinda: Okay, and when you say management, like you said with the Forest Service, part of it was like, they're just letting the fire burn the land?

Bob: Yep.

Brinda: Burn it up.

Bob: Especially the Wildernesses areas, they are not going in there harvesting anything. They're getting volunteers to do the trails, so they're not even having professional quality work done with the trail building.

Brinda: Okay. So you're . . . Would one of your recommendations for management be to actively thin some of these areas?

Bob: Oh, yeah.

Brinda: So that then fire could go through and not be as catastrophic?

Bob: Clearcut, thin, harvest. If you can, do it at a profit. Turn it into firewood. But you've got to keep managing the vegetation, which means you've got to keep disturbing the environment and that series of disturbances . . . What's good about the [National] Park Service is they say, "We want an 1850s pattern." So that's kind of . . . They have to do their research, and they have to do actions to get that, and then when they got it, they've got to maintain it. Well, I'd like to see the Park Service down here in the Biscuit [Fire] manage for it when they first had gold mines running in the Civil War era, 1850s or 1860s. Or even a little bit earlier, when they had Indians managing the land before white settlement, 1830s and 1840s. To me those would be wonderful times. Or even 1910. But right now we've got this unprecedented condition of tens of thousands of trees migrated in the area.

These invasive trees created contiguous ladder fuels, and they all burned up and then they burned all the old trees that had been there for hundreds of years with them, and of any spotted owls that followed, it burned them up, too . . . So, so much more for preserving the old-growth. Now it's firewood and it's all . . . "Now we got to keep them from logging."

Brinda: Okay. Well, okay. This is good. I mean --

Bob: I was probably getting way off on a tangent from your studies though. [Laughs]

Brinda: Well, you know a lot of the stuff that was . . . I think this is important still. In order for me to understand that broader context of what restoration means for different people even.

Bob: Yeah.

Brinda: And how it's being used because it's --

Bob: Political.

Brinda: It's political. It can be thrown out there and if someone doesn't ask . . . That's why I kept asking, so I could understand it from a different perspective. How is it being used? How are you using it? Because I heard it used in different ways. So it's really not a tangent in terms of my understanding it.

Bob: Yep, and it's a better way to understand reforestation to us. I think you're starting to be able to fine-tune the difference between the two.

Brinda: Right.

Bob: It's like, "reforest" is to put a forest back and it's almost always got a timber value component to it. Whereas "restoration" is to restore "functionality." I don't know what the hell it means, but there's almost always a government subsidy attached to whatever the project is.

Brinda: Okay.

Bob: I think that's a huge, huge difference.

Brinda: Yeah.

Brinda: And do you see a lot planting going on, on the fire . . . burned over lands. Don't they have to do it legally?

Bob: Well, on state lands, yes. Where we're getting oak restoration going now, that requires . . . For a while, it was still required that you couldn't do oak restoration because you had to plant these conifers within 24 months. Well, now salvage is against the law to do, and then plant. It's what they've done on Wilderness areas, I think is just a shame if we don't learn from it. I think it's been really counter-productive. If they're trying to save spotted owl habitat and old-growth timber, they've failed. If they're trying to save old meadows and berry patches and cultural landscapes, they've failed. If they're trying to keep management costs down by eliminating management, they've succeeded. It doesn't grow any taxes; it doesn't create any jobs. So, I think it's from a conservation standpoint, it's just wasted resources.

Brinda: What about for non-Wilderness areas? Say National Forests. Better not state land than federal land.

Bob: Right.

Brinda: Like the Biscuit burned 500 hundred thousand acres plus. Some of that was National Forest.

Bob: Well, most of it was National Forest. Yep.

Brinda: So then . . .

Bob: How much has been logged? Less than 7%, and that's creating all kinds of controversy. Out of the logging how much do they plan on planting with trees? So far about 100% of it, but people like me are throwing that into question. They haven't even got that far yet, and it's been three years.

Brinda: Right, because you're saying they can't plant until the stuff is taken out?

Bob: And they can't take a lot of it out because it's a riparian zone, because it's a Botanical Reserve, because it's a Wilderness, because it's an unroaded set-aside. It's like there's always a condition that you can't go in and harvest the material. So no matter if you go in there and plant, it's dangerous because you've got things hanging up there that are going to fall on you. We used to have crews out cutting those snags because they were dangerous. Same thing as a fire. If you've set it under controlled circumstances -- cut a snag and under controlled circumstances, it's dangerous, but not as dangerous as the one that just goes when it feels like it. Without warning and . . . so, and then even if you're going to plant it, those things are going to be falling on them and skidding down the hill. They're not going to do what the environmentalist say they're going to, which is to build the soils and protect the environment. They're going to cause erosion.

They're going to be wildfire traps. They're going to be a problem until they're managed. Even if the management is benign neglect and . . . They've got to accept the responsibility for the management actions or inactions and I think that's where the Forest Services has really failed miserably in the last 10 or 15 years. There's no accountability.

10. [1:42:54] "Environmental Restoration" & Race

Brinda: Okay, and maybe just the thing . . . You were saying that, in terms of the restoration movement, there were certain people, maybe key players, maybe not key players, but who got some of these projects going. [inaudible 01:43:16] restoration projects and these are graduate students or academics. So are there people that you have mind, or that I might read their stuff or follow up on? I mean, you said there's one professor you could think of.

Bob: Yeah. Let me think . . . If I can think of her name. I can find it out easy enough. She did it for about 10 or 12 years. She had an interesting name. So I can probably figure that out easy enough.

Brinda: Is she at or in --

Bob: At OSU. She used to be there for a lot of years. I don't know if she still is.

Brinda: Okay.

Bob: So there's her.

Brinda: And do they, would you say, pitch these ideas to the state?

Bob: Well, you'd probably go to the Oregon Natural Resources Council, or one of those groups. And they're the ones that lobbied for reduced cutting and then when that started turning up in fires, then they started pushing thinning and fuel management.

Brinda: But you're pushing for that too, right?

Bob: Yeah, or clear cutting. I'm just pushing for good land management. I don't care what it is, as long as it's science based, logical.

Brinda: Yeah.

Bob: Responsible.

Brinda: Right.

Bob: And hopefully self-perpetuated. So it'll produce an income so you can keep doing it rather than be dependent on tax revenues, so whenever Congress legislates, or the war on the right gets too big, or we have another hurricane, or something; it just doesn't stop.

Brinda: Okay.

Bob: And that's, I think, the fragility of having a politically developed industry replace an economically driven industry. I think that's . . . I'm not too sure if it's one of the things we've really seen.

Brinda: I like that quote. ["Uh-oh," Laughter] Okay, great. Right, because most of the outliers . . . And what I'm primarily looking at right now, like I said, is the public sector. Public lands and that work force and the change. The social history of change that is hand in hand with management practices, how that has affected that change. So, do you think some of this just won't be around in the rest 10, 15 years?

Bob: The restoration workforce?

Brinda: Yeah, and I don't even know what to call them. Whether to call them a restorationist, because they sound to me like --

Bob: That's pretty good. The environmental woods workers or you know, the restorationists is pretty good because it's the way they've used the word.

Brinda: Right but I'm thinking of the Hispanic workers part. Like say that was the [crosstalk 01:46:07] in the workforce.

Bob: I think they're the subset. I think they're the ones that actually do the work. There's a real racial element to the whole environmental movement.

Brinda: Absolutely. That's a big part of the thing.

Bob: Go into a Wilderness area and try to find a black family, or a Hispanic family. Or a really old person or an ill person or an infant. It's all these white yuppies. And they're not paying taxes on that land, they're already privileged. Why do they get to do what they want with it? What's wrong with rural people using it? So there is a real elitist, you know, "We don't log here -- we use logs from China and Canada" [Laughs] type of mentality hypocrisy that's going on that I don't think is sustainable.

Brinda: So what would you say about this racialized element of the workforce? Because the workers were doing a lot of what might be called grunt work. Whether it's limited amounts of planting still, or a lot of the thinning right now, they're spending all --

Bob: Or brush piling.

Brinda: Exactly. All of that, and it's really intense manual labor. I went out and saw these guys working, a lot of them, and --

Bob: Right. In this day and age, why are they doing that? Why don't we have black people down in the South cutting cotton by hand? Why don't we? We've got Mexican workers out here in Oregon hand-piling and where are they doing it? Federal land. Why aren't they doing it on private land?

Brinda: Are they not? Because of technology?

Bob: Because they're raking leaves. They're keeping busy doing jobs for . . . And there's supposed to be jobs to local communities. They're not supposed to be exported to a foreign country, but aren't we hiring foreigners that are sending the money home? And aren't we even doing in defiance of immigration laws? So why are we doing that? That's what I'm saying. There's an ethical element to the environmental movement, or the restoration movement, that's inherently racist and it manifests itself in lots of different ways. But it's privileged white people running it. And they're mostly lawyers and their well-educated kids that don't want to work, study or something, or want to get in the newspaper. I don't know what their motivations are. I don't think they're out for their children or grandchildren like they argued. A lot of them don't even have kids. My grandson is 15, and I don't think that's a true argument, and I don't they're conservationists. I think they're extremely wasteful and more out to be obstructionists than conservationists and when I think the . . . What they're doing with the Mexican workforce is not right.

Brinda: And do you think they're the ones controlling that workforce? Or is it [crosstalk 01:48:53] workforce. Is there a consequence of no one else wanting to do that work and --

Bob: No, we have plenty of people that can do that work. Nobody else wants to do that work illegally with no benefits for minimum wage.

Brinda: Right.

Bob: Just like nobody wants to be a brain surgeon for 12 thousand a year. Nobody wants to be a reforestation worker in which he's competing with the three Jose's who are here and there and maybe with the names of Frederico next week. It's like, well, who wants to be associated with that? Who wants to be 35 years old flipping hamburgers at McDonald's? You know, when they turn it into a minimum wage casual labor job, then there's no competition for that work because it's not real work. That's why I said before, it's more important that we're doing real work that needed to be done and had a consequence. It looked better, it was more productive, it was reproduction and people could see jobs, like whatever it was you were working for, emanating from that. From this stuff that's going on now, it's a little bit bizarre.

Brinda: So back then, okay? Back then, and I'm saying then, I'm talking about --

Bob: Late 60s to early 80s.

Brinda: Yeah.

Bob: Mid 60s to mid 80s is a real good, prettier time.

Brinda: Mid 60s to mid 80s.

Bob: Yep.

Brinda: So in my research it shows that there was a mix of crews and I'm talking about diversity in Oregon in terms of racial diversity to look at the racialized elements.

Bob: They developed. They even had Vietnamese crews for a while.

Brinda: Oh really?

Bob: Yep. The problem with those guys was they were so little. They were too little.

Brinda: Okay.

Bob: It was like, all of the equipment they had to use was too big and too heavy and they couldn't hack it.

Brinda: Okay, but there was a mixture of people out there once. There were Anglo crews. There were Hispanic crews.

Bob: Indian crews.

Brinda: There were Indian crews?

Bob: The Mexican crews . . . I don't know if Hispanic is exactly accurate. They're almost invariably Mexican.

Brinda: Mexican?

Bob: Like more of a national orientation. Because they'd be part Indian and part white.

Brinda: Okay.

Bob: But Spanish white rather than northern European white would be the main difference, and they had a lot of Indian blood in them. They developed, with the agricultural workers -- I think Jim Holt and just kind of an anomaly, and then as conditions became worse, they took over, basically. But they were like a real minority in the early 60s even until . . . late 60s or early 70s.

Brinda: Would you say that was that because the workers, the tree planters, it was that because they were getting higher wages and so there was a demand?

Bob: No, I would say it was because the feds favored casual labor. Because the feds would say, "Okay, we need somebody here three weeks." You could send in a group of Mexicans. Whereas my crew would say, "I want to go home to my wife at night. I'm not going to go over there for three weeks; I don't care how much you pay me. I've got a good job here." And so the private lands tended to develop stable localized workforces that tend to be white with a subset of Indians and a few Mexicans and the feds tended to develop workforces that were labor crews. Where they're Russian or Mexican.

Brinda: And now on private lands we have very little labor, your saying, because it's more technology oriented?

Bob: Yeah. That's part of it, and you get lots of Mexican crews because they're the only available ones that have the skills and that. Just like you don't hire timber falling crews anymore because of the technology. You can have a certain amount of machine planting but you're going to have to hire a contractor because that's what's been left over from this whole process that's running predominantly Mexican crews. Because, at this point, it's the most efficient. Basically what we saw was, gyppo logging industry, family-owned sawmill industry, and the reforestation industry, all dry up in the late 80s, early 90s. It's not just like reforestation got affected; gyppo logging outfits, the small sawmills aren't there either.

Brinda: Right. Well they were all kind of connected, right?

Bob: Yeah.

Brinda: I mean terms of a larger economy around --

Bob: First it was the loggers. Then the sawmills. Then the reforestation workers, which it kind of makes sense because you've got to log the trees off and you got to cut them up and then after a while you run out of ground and you need to start planting trees, so.

Brinda: Right. But there's some of the private lands that do run these crews. They run them because they're the only labor force now there that . . .

Bob: Yep.

Brinda: . . . is actually skilled and experienced in this.

Bob: Right. And also because of the transition, those outfits don't have the regular money, cash flow that they did before, and we don't.

Brinda: Which outfits?

Bob: Well, the private land owners and they don't have the idea, I guess would be it, that they're going to be owning the land 30 and 40 years from now. They're seeing themselves more of ephemeral land owners because that's the pattern. Rather than long term investors.

Brinda: Okay. So how are they managing their lands? Are they?

Bob: They sell them off and then somebody comes in and says, "I don't need to be paying these people to be working 30 thousand dollars a month to be doing this kind of work. I'll just hire one crew once a year for 60 thousand dollars and save 10 months labor and have them in and out and don't have to worry about."

Brinda: Right, and are you seeing just a, in general, as a private land owner, just people moving into real estate? I mean, selling the land off to developers.

Bob: A fair amount.

Brinda: There's more, in terms of more money to be made doing that than . . .?

Bob: And more demand.

Brinda: Yep.

Bob: Yep. A lot of that. Especially around the urban areas, not so much in the rural areas, but yeah.

11. [1:55:04] Private Landowners: Industrial & Generational

Brinda: And the private land owners. Who are the big land owners right now?

Bob: There's two groups. There's industrial land owners, and that would be Weyerhaeuser, Georgia Pacific, Boise Cascade.

Brinda: They sold it, right? Boise Cascade is now like --

Bob: Well, some of them sold it to themselves, then they call themselves The Timber Company and Plum Creek. I don't know if that's the Boise people or not, or Cascade Pacific. There's different owners that come and go and that they've incorporated. They're typically publicly dealt, and they own hundreds of thousands or millions of acres of timberland. And then there's the other ones. Some people call them -- oh, I hate this acronym -- oh, man . . . it's small, non-industrial, private forest; NIPFOs. Non-Industrial Private Forest Owners. But they're the individuals and families, and there's about 25 thousand of them; but the larger amount of timber is owned by the industrial owners and then the largest amount is owned in Oregon by the federal government.

Brinda: Okay. And that's what? 50%?

Bob: 50% of land but more importantly, 80% of the timber. So when they withhold timber from the sawmills, they're withholding most of the resources of Oregon. That would be like taking the corn out of Nebraska. Or 80% of it. Or the wine grapes out of California.

Brinda: But you were saying earlier that sometimes the private industrial timber land owners prefer that because that increases their value, sort of.

Bob: Yeah. The industrial land owners don't have to compete with the federal government. Industrial usually means you also own sawmills. So you're growing wood for yourself, and if you don't have to buy that wood at a high price from the federal government, you might charge a low price from yourself to make it more competitive, more apt to corner the market and that's just what they've done. The spotted owl has made a lot of money for a lot of private land owners.

Brinda: And at the same time put a lot of the cutters and a lot of guys out of work. Is that?

Bob: The rural economies have suffered.

Brinda: Well, right, because a lot of that money was being transferred to --

Bob: Jobs.

Brinda: More, and also weren't they giving ... Because it was so much public land in these rural areas, weren't ... I forgot the name of the thing now, but one of those things that funneled into the schools of --

Bob: Oh, yeah. There's two of them, yeah. There's . . . with the feds, because they weren't paying taxes, there's "Payments In Lieu of Taxes," which is the PILT money, but that's only when they were not logging but, let's see . . . it was the BLM gave . . . I've got to remember here, but they gave 75% of their net income to the counties. That's one of the reasons it's so popular, and BLM is only in western Oregon, really. It's nowhere else. Most of the federal timber for BLM is in western Oregon, and so they were building roads and schools and government buildings in Josephine County and that was a major share of their revenues for lots of years. The Forest Service wasn't that generous. But I think it was 25 or 50%.

Brinda: Percentage of that . . .

Bob: It was significant.

Brinda: . . . profits from timber?

Bob: Roads and schools. And when those ended, it was devastating for rural economies. But those again, those are kind of like the real loss is the jobs. The loss of family wage earner jobs and replacement with casual labor and no work. Because that's what keeps the communities together. It's the people. It's not the road or the school building. You've got to have parents there that are saving for college and [inaudible 01:59:22] backpacks somewhere, or whatever they do. And that was the part that was missing. It just turned into food stamps and classes and government subsidies. I don't think their people ever really caught onto to how humiliating and irritating that was for all the people that had work to do.

Brinda: And what happened to a lot of these loggers?

Bob: Well, you know, there's odd . . . There's kids who committed suicide. A lot of families broke up. A lot of alcoholism. A lot of . . . now there's people like me that went to college and our families dissolved and our businesses dissolved, and we went in another career direction, and there's some people that just hung on until retirement age.

Brinda: And this is where the reforestation crews as well.

Bob: Yeah. It's the same story with those three basic groups as far as the economics and what happened at the industries and why and where. They're all the same cousins and brothers and things like that, so . . .

Brinda: And did you ever go to any of these re-training programs? These Jobs In the Woods programs? Have you heard of those?

Bob: No. I looked at them and I had friends go for a day or so look at them.

Brinda: And what did they think of them?

Bob: It was where I was making comments earlier about graduate students [Laughs] and unemployed fishermen. That's what those programs were. They were a waste of taxpayers' money. There's real work to do. Nobody is doing it, so we got kids out in the woods fiddling around earning seven bucks an hour saying they're being retrained. Trained for what? Trained to do the very jobs that we couldn't find employment to do.

Brinda: Right, and a lot of the older folks, I mean let's say not think college age kids, but folks who are in their forties and fifties who used to be loggers, and they just didn't go back to anything.

Bob: Well, or they went to town and got a job roofing, or went back to school and got a job in the post office. A lot of them just flailed. A lot of the hardline rural people that have done it for a few generations and had a land base, pretty much just batten down the hatches that got them through the 30s so they knew how to get through the 80s and 90s. A lot of deer died. [Laughs] There was a lot of poaching and stuff like that.

Brinda: And did you see a lot of folks in the reforestation . . . tree planting workforce going with the Forest Service or the BLM?

Bob: None.

Brinda: Okay.

Bob: And I don't know any of them into restoration either.

Brinda: Okay.

Bob: It's a lot of them went fishing, commercial fishing, logging, landscaping. So related resource-based jobs and some, like me, education that have obviously stayed focused on the basic industry. I think people persevered in spite of the federal programs. The programs were insult to injury. They weren't helpful.

Brinda: When you say programs like Jobs In the Woods and stuff?

Bob: Yeah. That was just like a joke.

Brinda: Yeah.

Bob: It was like, "Wow, they don't have any money to do real work, but they've got all kinds of money to waste."

Brinda: Yeah.

Bob: And it's like everybody's out here having picnics or something or they think they're learning work and we . . . They're teaching the same types of jobs that one, we would be way better, way cheaper, we can't find any money for. But they'll spend 10 dollars to do a dollar's worth of work without debating whether it needs to be done or not, and then they'll all pretend like they're having an economic opportunity. They've taking people that never owned their own business and they're saying okay, we're going to create an independent contractor out of you. Well, if that person was ever going to be an independent contractor, they would have done it during the 70s or 80s when they had a chance. I don't know how many independent contractors have been created by federal government taxes but I'm guessing 0. [Laughs]

Brinda: Right.

Bob: And the contractors that they say, we got cars that need to be parked, or this or that. We've got services to do, sure. Then people will come up, but to be trained to be a entrepreneur, I don't think so.

Brinda: Right, but a lot of them, like you said, are coming up on their own because of . . . like fire, so you got a whole bunch of people coming in as independent contractors for that.

Bob: Right. Right. All you need is to have a policy of no logging and putting out fires once they start in the forest, and businesses will form. But that's getting back to the point of, why are we putting these fires out? If we're not going to log these trees, what the hell are we doing spending 130 million dollars on the Biscuit Fire? What are we trying to preserve? Yeah. That's where philosophically, I think that industry's pretty much shot its wad. I hope so. I thought so 10 years ago, and I was way wrong.

12. [2:04:43] The Environmental Industry

Brinda: Which industry?

Bob: The environmental industry. It's based on so much inefficiency, so many false assumptions, and hugely dependent on federal tax money. Might as well just be food stamps or something.

Brinda: What do you think of these Watershed Councils? Aren't there a whole bunch of these Watershed Councils that've been formed.

Bob: In Oregon.

Brinda: Who, I thought federal?

Bob: Feds support them. It's all federal money. They've gone through the State, yep. I think I mentioned them earlier, but I tend to have a caustic attitude on a lot of these things.

Brinda: That's fine.

Bob: Give me a different point of view. They already had the Soil Conservation Service, by river basins and drainages at a federal level. Now they've got it at state level, so it's what's called another layer of bureaucracy, in some places. I think they're another layer of bureaucracy. I think that the money they spend on those councils to get salmon back is like, everybody feels good, but they could all be dancing in circles around maypoles yelling for the salmon to come back and save a lot of money and have about the same result.

Brinda: Right.

Bob: I think it's government fueled, feel good, avoid the real problems kind of approach to salmon.

Brinda: This is not research related, but I'm just curious as to what do you think would be a solution, there's probably many for this, so the salmon numbers, which is I mean, objectively speaking and scientifically looking at it, one can say, salmon has decreased, right?

Bob: I don't know, I've done a lot of research on it.

Brinda: And you don't think so?

Bob: Well, coho salmon have increased a lot in the last 20 years. I fished for coho in 1982, and it was a terrible year. There are a lot of hatchery salmon, that are grown in place of natural hatches in the past, and that's the argument. It's hard to make a real quick answer on this, but a fairly quick one is, I saw a wildlife habitat map put out by the Wilderness Society, or one of those environmental groups, and they had these red areas of extensions, and they had veins next to Coastal Coho, and Clackamas Steelhead Run, and this and that.

They had a clear interpretation on them, it said all of these red areas had clearcuts. That's what they had in common. All these salmon runs don't exist anymore. That's what's called a . . . let's see, I'll think of it in a second. It's correlated, "spurious correlation." Because every one of the areas I went over with my friend, we just looked at it, here they spent, I don't know, maybe 150 thousand dollars and five years putting the map together, roughly, maybe more money than that, maybe a little bit longer.

We said, "Oh wait, that's the Long Tom Dam, that's Green Peter Dam" -- everywhere that had a red area, they'd have a giant concrete wall stopping the fish from coming up. Yes, they clearcut all those areas before they impounded them with water. The person who had looked at this clear cutting, saw these correlations, and hadn't finished their research; they went public with it. Were they ever embarrassed? No.

That's what had stopped it. They stopped the rivers, and they put in concrete walls. So, do we want to return salmon runs to those areas up above the dams? Well if we're going to, we'd probably need hatcheries or trucking or some kind of artificial maintenance. Or do we want to take the dams out and stop having electricity and irrigation. So, there's a tradeoff. We don't have elephants or sloths in the environment anymore, we don't have grizzly bears here in western Oregon. There used to be lots of them just 150 years ago.

With California Condors, we are trying to put some of them back, so it's the philosophical question of, if we want to keep salmon, we seem to know how to do it with hatcheries and fish farms and stuff, but if we want to return salmon to the spawning condition that they had 200 years ago, it's impossible. As soon as you take out the dams, now they can spawn up there, it's like okay, but now they've got to hang out in Downtown Portland or Seattle. Tell me what's the same about that? All those creeks got filled in. So, their habitat has been irrevocably changed.

Brinda: Right.

Bob: Whether they can persist in that habitat without human help, I would say they probably can, but it'd be in really diminished numbers. If we want them to persist so that we can have sport fishing and commercial fishing, it's pretty weird that we have coho, an endangered species, and a season on them. I don't know how much it costs to have each one of those killed by a commercial fisherman, or a sport fisherman, but it's a lot. Think if somebody took out a rifle and shot three spotted owls a year, it'd be crazy. Or a California Condor, it'd be . . . so it's political.

Brinda: Lots of lobbies. Okay.

13. [2:10:19] Dissertation: Forest Health & Human Health

Bob: Yep. That's probably more than enough once again, right?

Brinda: Right, I mean it's a lot, and it's so good, but I think it's sort of a lot to process.

Bob: Yep. Yep. Probably you'll find out you got way more stuff than you need for sure.

Brinda: Well, if I want to get finished with the dissertation, yes.

Bob: You gotta start cutting things.

Brinda: Yeah.

Bob: Have you started writing on it?

Brinda: I just started writing, and I have two draft chapters.

Bob: What's your hypothesis?

Brinda: Well, my hypothesis is that . . . well, it's sort of long and convoluted, but that the racialization of the workforce is related to changes in management practices. So it's very broad.

Bob: So it really is a lot of what we've been discussing.

Brinda: Yes, absolutely. Yeah.

Bob: Wow. That's interesting.

Brinda: The title of my dissertation, the draft title at least is . . . oh my gosh, I am so tired . . . I'll send it to you. It's not coming to me; it's one of those . . .

Bob: I can't remember my title and I worked on it for years.

Brinda: It's immigrant labor, immigrant labor and healthy forests; colon, something, something, something.

Bob: [crosstalk 02:11:47] deforestation.

Brinda: Healthy Forests is in quotes.

Bob: Okay.

Brinda: There will be a chapter on this whole healthy forests initiative.

Bob: I've got to give you one more quote then, is that still going?

Brinda: Yes.

Bob: I think a key difference, I think I said this on the phone already though, but I think it's a key difference, if we have to summarize everything; and I work with the healthy forest, healthy streams lingo a whole lot, that's why I get upset about some of these terms and take them personal. The key difference is, if it's got healthy kids playing in it, it's a healthy stream. The environmentalists will tell you, if you've got a kid playing in it, human health has got nothing to do with it. I think the same standards that measure human health by the . . . what's the whole union of all the countries in the world called?

Brinda: The United Nations?

Bob: The United Nations. Seems like that'd be something I can remember but I never deal with them. They have standards for what are healthy human populations, low infant mortality, educational standards, I think those are the exact same standards that work very well in ecosystem health, whether it's a forest or a watershed. Presence or absence of war? That's a wonderful thing. War really degrades the environment. Dying babies? That's always bad. That's usually something with mosquitoes or something. Poorly educated workforce, while poorly educated people do all kinds of environmentally disgusting things, compared to educated people.

There're all kinds of adverse effects from having an uneducated population. High unemployment, you've got the increased poverty and associated with that is harvesting of wildlife. Illegal harvesting and marketing, and so on. So I think there's the same standards for when you get into your healthy forest definition, I'm sure if you just use human standards and measure the human population within that forest or that ecosystem, you'll come up with a surprising result.

Brinda: In terms of the forest health?

Bob: Forest health is directly related to --

Brinda: Human health?

Bob: Human health.

Brinda: Yeah, so one of the things I'd --

Bob: Human population.

Brinda: Yeah. My title, I remembered, it's called *"Hidden and the Understory: Immigrant Labor and Healthy Forests in Southern Oregon."* It's sort of the last part of the story; you know but it's --

Bob: You need to talk to Jim Holt, then, and to John Foster.

Brinda: Okay.

Bob: Because you're in the southern Oregon area, and that's where the immigrant labor movement started, probably, in reforestation, and Jim Holt in Phoenix, Oregon is the first guy I knew that ran Mexican crews.

Brinda: Right. It's an interesting point you raised about human health standards, because an interesting take on it, and one of the arguments I might want to make. At least, there's this whole community forestry movement here; a lot of the folks who advocate restoration for local communities, you always hear this thing for healthy forests equals healthy communities, right? I've heard that a lot.

Bob: Yep.

Brinda: Those sorts of buzzwords. At the same time, there seems to be a hypocrisy there, because it's talked about, but then, the people doing this work are often, at least now, are often highly vulnerable, or highly exploited. A lot of that is because they're here without papers. One of my tongue-in-cheek things is, are we maintaining healthy forests, but are they really healthy communities? People aren't being talked about.

Bob: That's not tongue in cheek, it's the problem; it's the elephant in the living room exactly like --

Brinda: It's happening on federal lands, so how is the Forest Service getting those on the cheap, and how is that happening and no one talks about it, no one talks about immigrant welfare --

Bob: Except it's not cheap. They're getting the labor cheap.

Brinda: They're getting the labor cheap.

Bob: But the labor they're doing, is costing everybody a lot of money. It's not labor that pays for itself; it's getting back to the busywork problem. It's not cheap. The only ones that are getting screwed over --

Brinda: Are the workers.

Bob: Are the workers, and the communities that these workers are living in. Think about it: if the important thing is to keep a hygienic kitchen to reduce human health problems, and the way you do that is you have an illegal Nigerian lady come in, and spread Lysol everywhere, that's not a very healthy environment. That's the exact point you're getting at is the exact point I've been trying to make. They're controversial points, and they're important points. We're subjugating entire groups of people based on race. It doesn't matter if they're dead Indian people, or living Mexican people, which usually have more Indian blood in them than our American Indian populations. There's this whole privileged, white perspective on Wildernesses, and ecosystems, and healthy forests, and restoration that permeates the whole industry. The only way they're able to get away with it is federal laws, and federal land ownership.

You're seeing a pattern, sounds like you're getting it there, the feds are pretty much the ones creating this situation.

Brinda: I'm walking on very controversial ground --

Bob: Yep.

Brinda: And sort of dicey ground, because I'm revealing my cards here, and I'm just trying to go with the data, what it says, presents that.

Bob: Yep.

Brinda: I'm not necessarily saying these workers are terrible and need to leave.

Bob: We had a bunch of . . . I hired a lot of Mexicans on my crew.

Brinda: I believe that I'm researching people who are contributing to the economy, paying your taxes, but because they don't have their papers . . . many of them have lived here for many, many years, and are no longer migrant workers. That's part of the story, I've got a chapter that's looking at exactly what this interview has been about, and our interview over the phone is sort of looking from the period from the 50s to the 80s. That's looking at the ARC [Associated Reforestation Workers]; that's looking at a more Anglo-based workforce, and how it was organized differently and how it started changing.

Chapter three and chapter four is going to look at more my interviews with the Latino population, and some of the workers, and those contractors, and there's a lot of internal exploitation going on there too.

Bob: Do you speak Spanish?

Brinda: I can get by.

Bob: Oh, okay.

Brinda: I also worked with a Latina RCC student, who came around and learned how to do interviews and got really passionate about this stuff.

Bob: Oh, good. Good.

Brinda: It's an interesting question, it's not necessarily saying, exposing that workforce and saying, they're terrible they have to leave, but just showing like, what are these dynamics that ends up with this highly racialized and vulnerable work force.

Bob: Looking at the workforce and saying it's terrible how they're treated. That's the bottom line.

Brinda: Looking at that. The bottom line is looking at that, but then also looking at sort of how environmentalists don't even talk about labor issues. They just talk about the trees, okay well . . .

Bob: "Naturally functioning ecosystems."

Brinda: It's just like how a lot of this takes place on federal land.

Bob: Most of it takes places on federal land.

Brinda: Just like, exposing this thing. People talk about farm work, they think automatically Hispanic labor, Latino labor. When we talk about forests, people think . . . well they might of loggers, but they think about the trees, it's just been represented differently.

Bob: It's all Caucasians.

Brinda: But it's not.

Bob: If they're looking at recreational use -- I know . . .

Brinda: Yeah, yeah.

Bob: When they're talking about recreational use, they're right.

Brinda: Yeah.

14. [2:19:45] Mexican Labor & Environmental Blinders

Bob: When they're talking about the labor, it's just like you and I are talking. There's some real transitions, and they're racial transitions. What sets us between Mexico, or Hispanics, it's kind of like Native Americans . . .

Brinda: It is Mexican. It is primarily absolutely Mexican.

Bob: It's like, I don't even know what a Hispanic is exactly. That could be an American with . . . so we've got these politically correct terms, but it's still a racial issue. Again, it's an Indian issue. The people in Mexico, of Indian heritage, are the ones that are working on the farms and coming up here.

Brinda: A lot of them.

Bob: It's not the Spanish people that are controlling the politics, and the economy. It's again, Indian people. It's the secondary racial group of Mexico that we're talking about.

Brinda: Yeah, in Mexico.

Bob: Yup.

Brinda: It's just a really interesting, fascinating issue. The thing is, once this word gets out there, right, you have no control ultimately of how it will be interpreted.

Bob: That's the scary part for you.

Brinda: Yeah, it is.

Bob: A lot of times I get interpreted as a total racist, because I'm talking about racial issues, and I'm supposed to be politically correct, because a middle-aged white person is just supposed to be . . . just like you're talking about the environmental movement, they're not even talking about it. It's institutionalized. It's so deeply embedded in them, that when you bring the topic up, it's like, repulsive.

Brinda: Or people don't want to think; you know, they're not that way.

Bob: That's the repulsive part of it. It's like, "I'm not a racist, this isn't happening, how could that be happening? It's got to be an abhorrent situation, because the forest as I know of it is a mushroom, growing out of rotting wood kind of thing," that they and their friends go out and look at in nature and photograph. But they don't see the guys out there actually piling brush, which are convicts or Mexicans.

Brinda: Right. Right, and looking at that and exposing that workforce of, once this word gets out it will be interpreted one way by an Environmental Movement, and it will definitely be interpreted by some more, maybe conservative anti-immigrant elements to society, and a way to say these people need to leave. There's all kinds of things that could happen.

Bob: A lot of it's going to depend on, even though what they say on their dissertation is, recommendations and a dollar is enough for a cup of coffee, maybe, but the recommendation to me are the most important part. They'll say it's the hard data, and demonstrating that your data's there, and it's clean, and you've organized it and you've done your job as a scientist. That's really important to do that. Then there will be a lot of people who will argue, that's the main thing you can do as a graduate student, is the data.

Brinda: Right.

Bob: But, I really think it's the insights that when you take the data, and work with it creatively, and you come to conclusions or recommendations, that's the part that's going to give you power.

Brinda: Right.

Bob: It's one thing to have the data and just say, well everybody just take it. That lack of accountability is the same problem that the Forest Service has.

Brinda: Yeah.

Bob: If you're accountable for it, you can make the argument right from the [Unintelligible 2:22:53]. These people are doing valuable work or wasting their time doing busy work.

Brinda: Right.

Bob: If they're doing valuable work, they need to be treated like a valuable commodity. Just the fact that you're even talking about these people is like, a shock to those people. It's like, who are these people? That's your rural jobs, that's your local economy, that's your diverse work force that you're all arguing for, you just don't want to look at what it really means.

Brinda: Well, I can see who's in your community. A lot of the local communities you live in, maybe rural Oregon, but a lot of these --

Bob: I'm an academic --

Brinda: -- the composition of these communities have changed a lot, into what is local.

Bob: Yep.

Brinda: Local is settled life, but it's also still Mexicans who are both contributing to that economy, and that's not necessarily voiced by . . . I wouldn't say it's the local people who are sort of against each other, it's more that there's this [environmental] "movement" talk that's out there.

Bob: It's also urban fueled.

Brinda: Yeah.

Bob: It's like privileged people that probably aren't even aware of their privilege. They believe the voice of the underdog is crap; they're the voice of privilege.

Brinda: Well, you know, anyway. So that's kind of what I'm ultimately showing, is sort of that social history --

15. [2:24:12] Ecology, Sociology, Race & Economics

Bob: What's your associate, that's working on the restoration stuff?

Brinda: Well she's not, her research . . . she works for the same advisor I do.

Bob: Oh, okay.

Brinda: Her dissertation topic is . . . so we're not a tag team, but it'll be interesting to see what comes. She's looking at restoration itself. The restoration of a forest, and how a lot of the restoration industry is . . . the projects themselves are not successful, it'll be like doing . . . some of the crazy things she's done, these interviews, all this money goes into sort of creating these stream beds and stuff, but there's no water. It's like a stream bed without any water.

Bob: I'm dealing with that right now. They've got big riparian zone, it's like . . . riparian means water, there's no water there, but they're saving all this vegetation because it's ephemeral: sometimes there is water there.

Brinda: Okay, well I don't even know if there's any water in this place, she's just saying how this whole thing has been created trying to be this industry, but . . . even in terms of environmental standards or ecological standards, it doesn't seem to even make sense.

Bob: It's not scientific.

Brinda: Who is this work -- a lot of these people are more like an Anglo-workforce, she just doesn't see . . . her argument --

Bob: It sounds like she's coming out with a similar viewpoint as yours or mine its --

Brinda: Yeah, well she's saying the way the restoration workforce identifies itself, I think she's saying it's also racialized, but it's primarily through the middle class white people who are saying, we're doing high skilled restoration work, they're not even talking about Mexican workers.

Bob: Yep.

Brinda: She's also saying it's sort of an industry that tries to fuel itself.

Bob: Yep.

Brinda: It's really not doing anything. She did her research up in Ukiah, near Ukiah in Northern California.

Bob: Good, good. Yup.

Brinda: You know, near Hopland, that area.

Bob: Probably there's a "First American . . ." oh, I can't think of it . . . I just found it real [crosstalk 02:26:14]

Brinda: She interviewed Watershed Councils, people doing this stuff. She's more, again, social science-y, she's looking at the narratives of restoration, how do people talk about this thing, and what does it actually mean? It'll be interesting what comes out of it.

Bob: When's she going to get it done?

Brinda: Hopefully, this summer, too.

Bob: So you're both people who can state your goals and objectives and you can actually finish them this summer, or you're --

Brinda: Oh, I think she will. I hope I can too. I have to, because I've run out of funding, and so I need to be done and find . . .

Bob: I ran out of funding, that didn't get me finished.

Brinda: Well, I need to be finished. I think she'll be done by December at the latest.

Bob: Oh, wow.

Brinda: She's tired of being a grad student as well.

Bob: You two both plan on being teachers, or researchers, or . . .

Brinda: I don't know what she wants to do, I think she might want to go into . . . you know, she's not sure yet, she's pretty young, she wants to do policy work, maybe. I wouldn't mind doing something at a community college.

Bob: Wow.

Brinda: Or teaching. I went to the Forest Service because I just moved down to Riverside, California. My husband got a job there. When I went to the Forest Service, they have a Research Station down there, I was really just kind of disheartened by some of the research they were doing in terms of . . . they were just doing surveys. Social science surveys of how people interpret signs in the forest, I'm not really . . . well, we'll see I don't --

Bob: I'm putting out proposals for the Pacific Northwest Research Station to Oregon State, just so I can get a reasonable project together. Because, they've had their funding cut back so much that they've just turned into --

Brinda: The Research Station?

Bob: Yeah, all the Research Stations, they won't let the Forest Service do any research. The Forest Service doesn't have many to monitor, they'll make the big, massive, expensive decisions but only in context of if they're fighting a wildfire or something.

Brinda: Yeah.

Bob: It's a little bit bizarre, but I'm really glad you're doing what you're doing, and I'm glad your friend's doing what she's doing.

Brinda: Yeah.

Bob: You're coming up with really critical insights --

Brinda: Right.

Bob: -- that you're finding out aren't shared by hardly anybody. Even people that are right next to the ground that are in the Environmental Movement stuff, are not sharing your perspective.

Brinda: It's a good thing to talk to you about that, because when we get in our little critical thing, I start thinking like, everyone must see this. The whole restoration thing, I was saying, well aren't people seeing that -- and you're like no, they're saying something completely different. As you're thinking critically and analyzing this, I get these sort of blinders on, not blinders but you start thinking well, isn't this evident? Everyone must see that it's there, but they don't, and I forget that.

Bob: The one thing I've found with the Environmental Movement, is that it's a lot of urban kids, most of them have come into in the last 10 or 15 years. They've haven't grown up there, multi-generations, they really don't understand the workforce, they don't even understand the context we're talking in, who's actually doing the work, how's that related to what kind of policy --

Brinda: Right.

Bob: -- created that kind of job. There's a self-assurance there, and they just out of things. I've been to a couple public meetings at the last [inaudible 02:29:29] so this isn't . . . recent experience reaffirms, it's not only do they act like they know more than they really know about ecology or the history or economics; they act like, a lot of them, like they know more than is humanly possible to know. It's like, all of a sudden they're talking about fungus, and they're talking about labor things, and they're talking about fire suppression, and it's a little bit annoying. It's like they're really brilliant people; except they're just [inaudible 2:30:01] to me. That's been kind of a problem I've had. I used to think they were well intentioned, but they just needed better information, then I came to the conclusion that's not it. They reject better information, they're pretty much self-serving people that have their own agendas, that have a combined vision or viewpoint that is common to them all, but they might as well Rajneeshees, or I don't know, Mormons on a mission or something. They've got something driving them -

Brinda: Right, I mean --

Bob: It's not history, and it's not logic, and it's not resource management, and it's not economics.

Brinda: Yeah.

Bob: There's a large dose of racism there, and they're just oblivious to it.

Brinda: Right. Well thank you for that fantastic interview, which I'm going to have to listen to again and again --

Bob: Uh oh, I'm sorry.

Brinda: No, no, it'll be great.

Bob: Only because you're interested!

Figure 2.8 "Crew of the Month." Undocumented Latino forest workers being arrested by Immigration and Naturalization Service officer. Siuslaw National Forest, Waldport Ranger District, OR. Source: *Newport News Times* staff photo, reprinted in *ARC Quarterly* (Winter 1982): 19. (Sarathy 2012: 44).

Forest Restoration: Problems and Opportunities

By Bob Zybach, PhD.

Actively managing our western forestlands on a landscape-scale can immediately create thousands of rural jobs, greatly reduce catastrophic wildfire risks and damages, return millions of dollars to our state and federal treasuries, increase native wildlife populations, fund our rural schools, roads, and libraries, and make our forests and grasslands safer and more beautiful than ever before. **Seriously.**

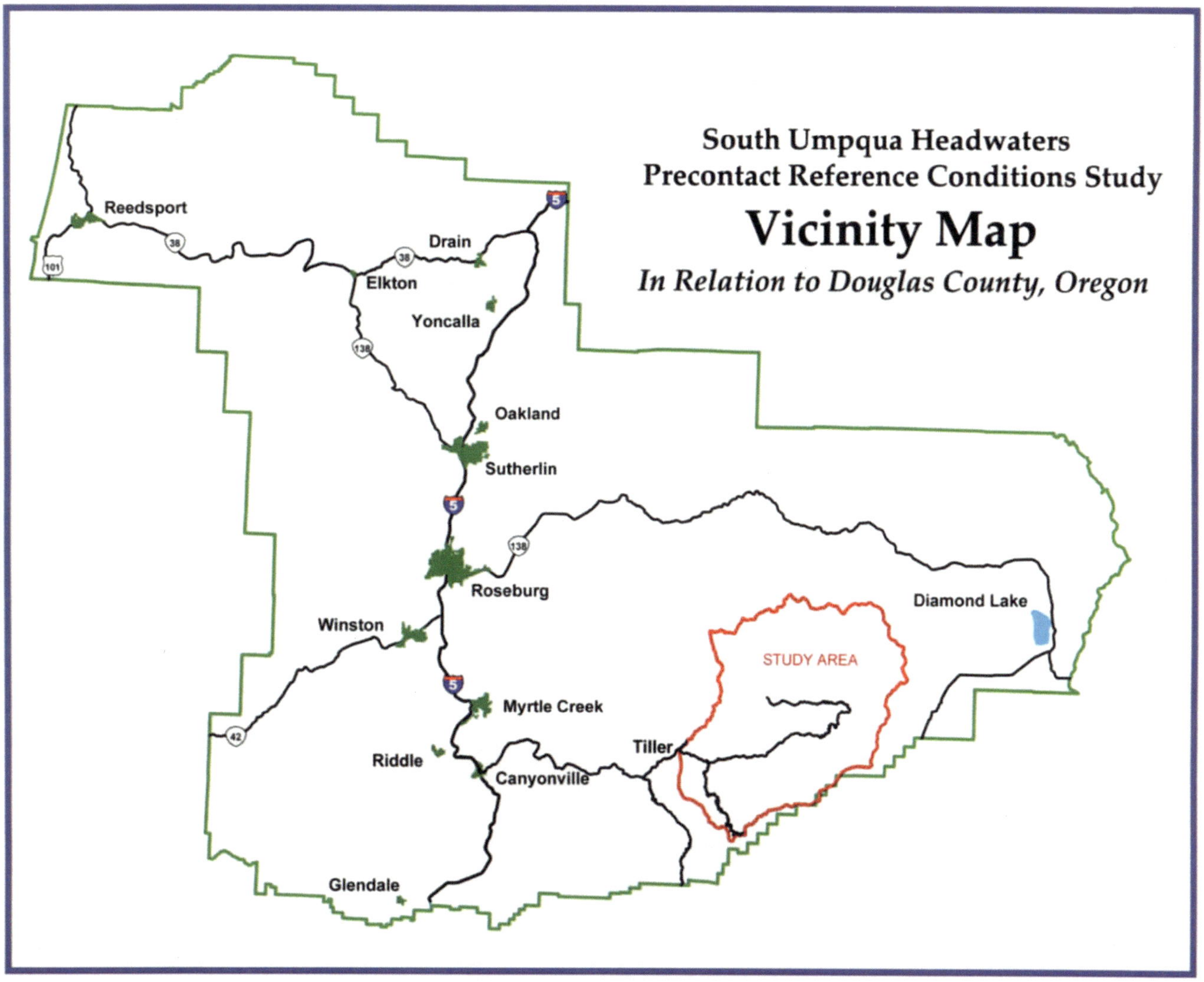

Western forestlands have never been in worse shape: millions of acres of dead and rotting trees; thousands of miles of abandoned and barely maintained roads; record wildfires becoming larger, deadlier, and more destructive by the year; hundreds of artificially impoverished rural communities; and endless litigation preventing the use of resources we need to sustain our lives and our economy.

There are a number of reasonable ways to resolve these problems; a long-term commitment to active forest restoration and management seems to offer the most immediate benefits to both people and wildlife, and is the likely route to economic sustainability as well.

What is forest restoration, why is it needed, and how is it done are the questions addressed in this article. Two

Map: Forest Type and Trail Patterns, 1800-1825

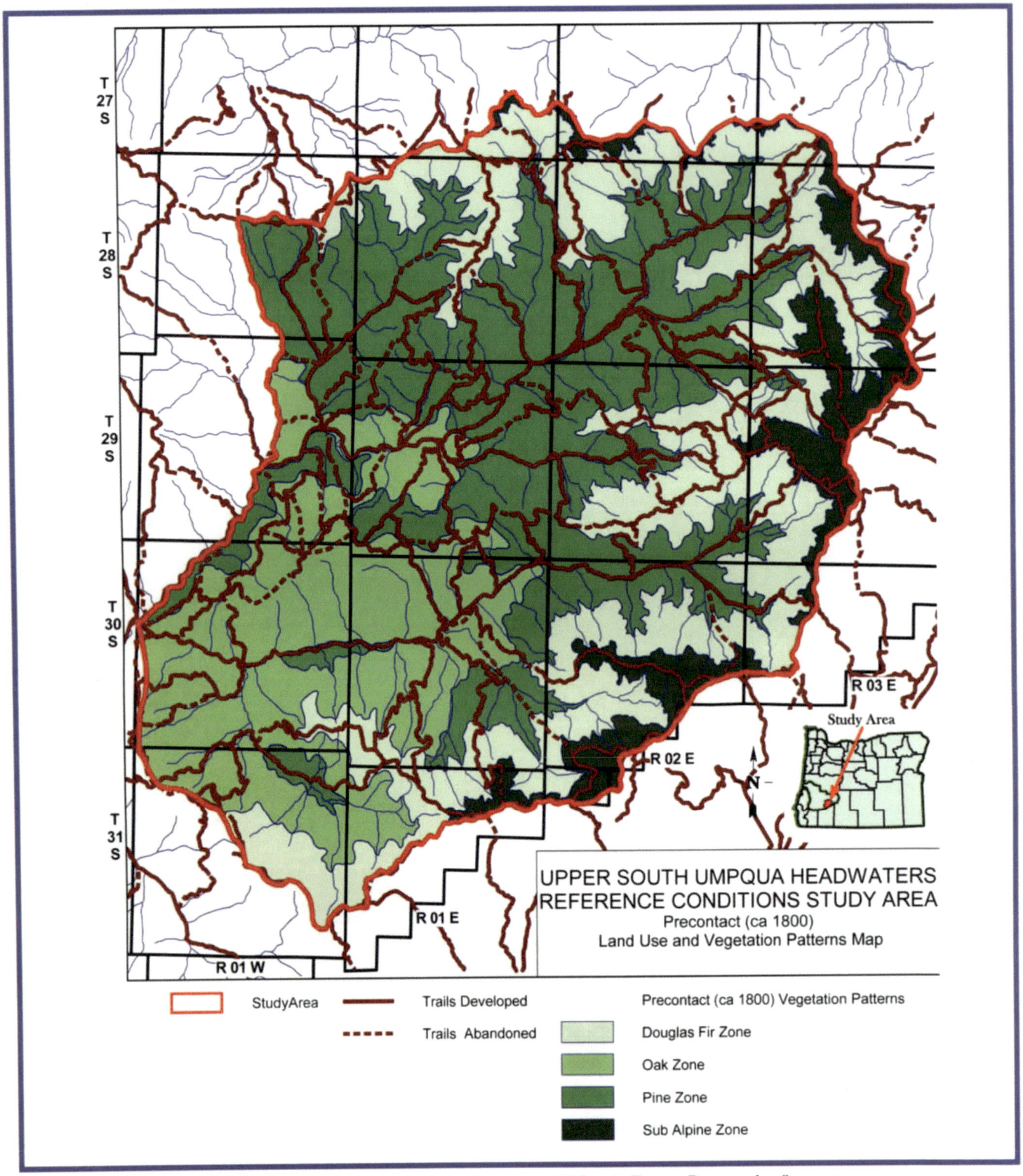

examples of current forest restoration projects are profiled to help answer these questions, and illustrate how these types of programs can be immediately implemented across the landscape to the benefit of neglected forests and depressed timber-dependent communities throughout the West.

What is Forest Restoration?

The process of forest restoration is focused on returning an area to one reflecting desired past conditions, it is critical to understand a) what conditions were actually like in the past, and b) which of those characteristics (if any) should be restored or preserved for the future.

For the past 10,000 years and longer, Native Oregonians have used plants and animals for their own purposes: principally for food; shelter; fuel; and fiber products, such as clothing, basketry, musical instruments, canoes, ropes, and weapons. Fire was used for a wide range of purposes: for cooking, heating, and lighting areas around homes and campgrounds; for rejuvenating berry patches and harvesting fields of grain; for hunting game by systematically setting vast tracts of land on fire.

Photograph 1: Oak Type. Former oak and pine savanna.

Man is the only animal that can use fire, but he is not the only animal that benefits from it. The expert and judicious use of fire across the ancient landscapes of Oregon resulted in the stable patterns of forests, woodlands, vast prairies, wetland meadows, breaks, balds and berry patches encountered by Oregon Trail immigrants in the 1840s and 1850s. Elk, deer, songbirds, fish, squirrels, migratory fowl, and other animals that populated these environments were documented by many of the new residents.

Forest restoration, means restoring people to the land, whether in the woods, along a river, or walking through a town. Restoring people to the land also supposes restoring fire to the land; fires set by people, not lightning.

Upper South Umpqua Project: Considering Past Conditions is Step 1.

The map shown in this article represents a critical step in the forest restoration process – a determination and documentation of likely past conditions for areas being considered for restoration. Whenever we plan to restore something, it is important we understand the conditions that existed in the past.

The Upper South Umpqua Headwaters Precontact Reference Conditions Study focused on characterizing a significant portion of the Umpqua National Forest in Douglas County, as it likely existed in 1825. The study area is slightly more than 230,000-acres in size and extends from the crest of the Cascade Range westward to the confluence of Jackson Creek with the South Umpqua River. The map shows the location and composition of forest type patterns as they likely existed in the study area 200 years ago. Each of the subsequent four photographs documents a typical example of each of the four forest types, and illustrates potential forest management actions needed to restore and maintain desired future conditions.

Photograph 2: Pine Type. Invasive Doug-fir/madrone.

One of the basic purposes of forest restoration is to reduce wildfire risk and damages. The method for achieving this in overstocked stands of conifers is to significantly reduce their biomass ("fuel load") and open up the tree canopies ("thinning") as they existed in earlier times, when catastrophic-scale crown fires were uncommon occurrences. On federal lands this is referred to as an "FRCC 1" condition.

The Upper South Umpqua Project was initiated by Douglas County Commissioner, Joe Laurance, to consider the possibility of restoring degraded local forestlands to presettlement condition. On July 15, 2010, he testified to a Congressional subcommittee of The House Natural Resources Committee in Washington, DC:

Fire Regime Condition Class (FRCC) 1 is similar to the forest which European explorers first found here. That forest had been modified by fire for more than six thousand years to provide the native inhabitants with what were then life's necessities. These included abundant wild game from the most productive and diverse wildlife habitat ever known on this continent. Similarly, the regular burning of competing vegetation permitted propagation of nut bearing trees and other food producing plants. Additionally, the historic "Healthy Forest" promoted pristine rivers, streams, and lakes that provided an abundant harvest of fish and waterfowl. Within FRCC 1 the risk of losing key ecosystem components to fire is low, while vegetation species composition, structure, and pattern are intact and functioning within the natural historic range.

Research methods used to determine and document 1825-era forest conditions in the study area included extensive use of General Land Office survey maps and notes, historical maps and photographs, field plots, oral history interviews, literature reviews, archival research, and over 5,000 GPS-referenced digital photographs. This latter method documented the location and extent of remaining old-growth (pre-1825) trees in the study area, in addition to documenting persistent patterns and patches of such traditional cultural food and fiber plants as camas, fawn lilies, cat's ears, huckleberries, hazelnuts, chinquapin, tarweed, serviceberry, wokas, bracken fern, thimbleberries, and salal.

Historical research has given us the map shown: a generalized depiction of likely forest conditions in the study area during the 1800-1825 time period. The following four photographs represent current typical conditions within each of the four forest types (or "zones") shown on Map 4. The large size and wide spacing of the older trees in the photographs can be gauged by the "human scale" used to measure them: Nana Lapham, long-time forest science research assistant of NW Maps Co., did much of the field work on this project.

continuous canopy of fine, pitchy fuels. A crown fire under these circumstances would likely kill all of the trees in this picture, including the old-growth.

Photograph 2 is a typical condition found throughout the former Pine Type in the study area. Again, these areas were maintained by regular prescribed fires in precontact time, and

Photograph 3: Doug-fir Type. Douglas-fir/pine mix.

are now being threatened by thousands of invasive Douglas-firs and madrone.

Photograph 1 shows relict trees from an oak and pine savanna that was developed and used by native residents for hundreds or thousands of years. These areas were tended by constant gathering of fuel, acorns, and pine nuts from the trees and regularly tilling and/or burning the surface area to manage understory crops, such as camas, tarweed, hazel, and beargrass. Notice how Douglas-fir have invaded the area in the past 100+ years, threatening the survival of the few remaining old-growth oak and pine.

Restoring this savanna would entail removing the Douglas-fir, before they smother the remaining old-growth, removing the surface fuels that have built up around the bases of the oaks and pines, and reintroducing the understory plants and regular burning practices that created and maintained savanna conditions in the first place. There are thousands of acres similar to this throughout the study area, with invasive Douglas-fir slowly killing the established old-growth and creating potential crown-fire conditions by developing a

Photograph 3 shows the 1825 Douglas-fir Type, which still contains scattered old-growth sugar and ponderosa pine; species which may have dominated this type in the Indian era. Restoration provides the best option for prolonging the lives of these historic trees, too.

Photograph 4 is a picture of Huckleberry Lake, in the high elevation True Fir Type. The "lake" is now a wetland prairie. This area used to be a portion of a major huckleberry gathering complex, equally accessible to Indian families living in the Rogue River and South Umpqua basins before these plants were shaded out by the invading conifers. The taller trees in the background represent larger diameter, older trees likely dating to the 1600s and 1700s; the vast majority of trees, though, having established themselves after the mid-1800s and throughout the 1900s.

The common theme throughout this map and photographs

is that tens of thousands of acres of old-growth oak, pine, Douglas-fir, red cedar, and other conifers exist throughout the study area are in need of immediate attention, if they are to survive. The same problem exists for the scattered patches of huckleberry, camas, tarweed, and native grasses that still persist. A forest is, ideally, composed of many facets, housing many different types and species of plants and animals. Those are the types of attributes that used to define these lands, and the same types that can be restored and maintained for future generations of people and wildlife.

Photograph 4: True Fir Type. Huckleberry Lake.

Jims Creek Project:

Make a Choice and Then Do It is Step 2.

The second step to forest restoration, is to determine what future conditions are desired and to begin actively restoring and/or maintaining those desired conditions across the land. An excellent example of this step is the Jims Creek forest restoration project on the Middle Fork Willamette Ranger District, which is in the process of completing an initial 400-acre "demonstration project" portion of a 25,000-acre plan.

The Jims Creek project has demonstrated the feasibility, profitability, and general benefits of conducting landscape-scale forest restoration projects on federal forestlands, but it is also a good example of how much time can be spent in putting these projects together, as well as the ease and quickness with which they can be stopped by adversarial legal actions. This project was initially conceived by a local US Forest Service forester/project manager, Tim Bailey, who then spent the better portion of the next ten years shepherding his vision through the myriad public meetings, scientific reviews, committee presentations, promotional tours, and other hurdles needed to get things underway on the ground.

Picture 5 shows Bailey in front of a portion of the Jims Creek Project in 2010. This area had already been treated by removing most of the invasive conifers established during the past century, and by broadcast burning the ground so as to remove excess litter and logging debris. Note the scattered trees that have been left behind: widely spaced pine of several different age groups, from seedlings to saplings and second-growth to old-growth. Also note the small herd of elk grazing directly above Tim, near the crest of the hill.

Picture 6 is the same herd of elk as the previous picture, seen through a zoom lens. The small charred stumps and large woody debris in the foreground will soon rot away or be consumed in the next few surface fires. The reddish-brown pattern is bracken fern, a plant harvested in large quantities by many Oregon tribes for its starchy roots and asparagus-like "fiddleheads" that grow in the spring. In a few more years, with a few more broadcast burns, this area will appear very similar to what it must have looked like hundreds of years ago.

Additional restoration has been halted at this time due to an infestation of red tree voles ("tree mice") that accompanied

Photograph 5: Tim Bailey and elk herd on Jims Creek site.

the migration of Douglas-fir trees into the project area during the past century. These rodents are protected against logging under a federal "survey and manage" regulation, despite the fact they are not a threatened or endangered species.

This type of work stoppage, based on new federal regulations and litigation initiated by environmental organizations, has become the main difficulty in beginning and completing forest restoration projects in Oregon and throughout the West. Of the 25,000 total acres of this project within the Middle Fork Ranger District, 7,000 acres are privately owned and managed for maximum timber production; more than 7,000 acres have been classified as spotted owl habitat; approximately 7,000 acres are in remote areas that would likely include the "taking" of spotted owls; and the remaining 4,000 acres are populated with regulated tree voles. A river flows through the project area that contains two listed fish species and "there is a perception on the part of the regulatory agencies that this type of restoration work can have a negative effect on fish."

Despite these hurdles, there is a lot of work needing to be done and a lot of people wanting to do it.

Conclusions

Forest restoration projects should be conducted on a landscape-scale basis in order to be effective biologically, aesthetically, and economically. Project boundaries should include sufficient commercial materials to treat the project area and show a profit. Profitable and beneficial actions are sustainable on a long-term basis, as we have learned from more than 10,000 years of forest history in this region.

The actions needed to restore our forests to earlier, more desirable conditions would create thousands of jobs for decades, jobs to make the best use of our resources, protect our old-growth and its wildlife, and greatly reduce the likelihood and severity of wildfire when they did take place.

Based on my own experiences and observations, I think there are four things that must be in place for forest restoration projects to be successful on a long-term basis:

Photograph 6: Close-up of 15-16 elk in restored prairie.

1) Areas slated for restoration should include sufficiently broad boundaries and specifications to allow projects to be profitable;

2) Restoration projects should be landscape-scale (25,000 to 250,000 acres) in size in order to be economically efficient and biologically effective over time;

3) Local residents and businesses should be in strong support of restoration projects, and be given access to all information that develops during the process;

4) Local project managers should be knowledgeable and capable of communicating scientific, technical, and political aspects of a project to local citizens.

I remain certain that the adoption of these practices, as defined, would have many immediate and positive effects on forest health, old-growth preservation, endangered species protection, rural economies, international trade balances, and many other economical, ecological, cultural, historical, aesthetic and recreational values associated with Oregon's forests.

The degradation, destruction, and loss of our federal forests and grasslands to wildfire, bugs, and disease will continue to escalate so long as we continue our current path of passive avoidance and neglect. Restoring our Nation's forests means restoring people – in part, as active managers – to our lands. The benefits for doing so have been listed; the impediments to getting started have been largely self-inflicted, are almost entirely political (rather than scientific or humanitarian), and can be readily surmounted, given effective leadership or common outcry. The best time for doing something

Pacific Northwest Forests:
A History of Disturbing Events and Circumstances
By Dr. Bob Zybach

In late January I joined with recently retired Starker Forests forester Dick Powell to lead a seminar on forest history for the annual 2015 Pacific Northwest Forestry Leadership Academy held at Oregon Garden, near Silverton, and sponsored by various forestry organizations in the state.

History (1)

Fig. 1. Drawing and written description of a condor at Fort Astoria by Meriwether Lewis, Feb. 16, 1806.

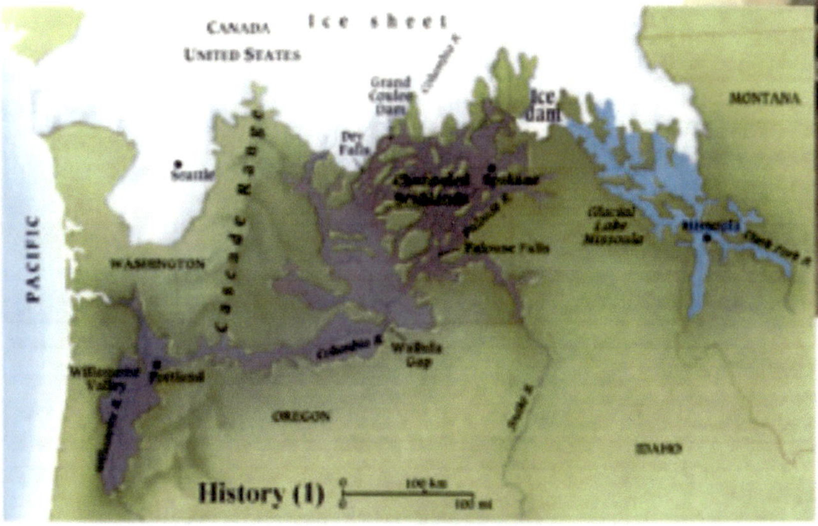

Fig. 2. Map of glacial extent and Bretz flood events 12,800 to 15,000 years ago in the Pacific Northwest.

Fig. 3. Glender Brothers elephant tooth taken from their Benton County, Oregon spring, circa 1919.

Dick gave the first presentation, and concentrated on the history of forests and forest product uses throughout the world, and particularly as they related to the evolution of western civilization. His article, summarizing his perspective precedes mine in this magazine. My presentation, intended to complement Dick's, focuses more specifically on the forests of the Pacific Northwest, from the last ice age to the present.

What is "History?"

There are two basic definitions of history: 1) the period of time in which local events and conditions have been documented by eyewitness accounts via written words, maps, drawings, photographs, videos and/or other common methods, and 2) the period of time in which people have been present in a given area.

Figures 1 through 3 illustrate these two definitions, showing environmental conditions that existed during early human occupation of the Pacific Northwest: eyewitness documentation of a currently endangered bird species that freely existed in the region 200 years ago; and a photograph of a man who had discovered an elephant tooth in his spring in western Oregon – left from when elephants, condors, and

Fig. 4. Wind Mountain landslide and submerged forest, described by Lewis and Clark in 1805.

Fig. 5. 1894 landslide creating Gould Lake.

people all coexisted across the regional landscape.

Recent archaeological finds have placed people in the Pacific Northwest more than 15,000 years ago, a time when major glacial retreats and biblical-scale flooding and landslides were caused by the melting ice and rushing floodwaters. As glaciers and waters receded, it is very likely that human footprints preceded the first tree seeds and subsequent forests in those areas of northern Washington, the high Cascades and the Columbia and Willamette valleys that became forested during historical time. Some people have described Pacific Northwest forests as "ancient." If that is true, then it must be recognized that people in this region must be more ancient still.

Circumstances vs. Events

Circumstances can be defined as a set of conditions, facts and other details that can determine the likelihood, extent or severity of an event -- and that will likely be changed or even eliminated as a result of such occurrences.

One example is the concept of "wildlife habitat" that has been developed in recent times to describe the set of circumstances that some scientists claim provide optimum forest conditions for select plant and animal species, such as spotted owls, coho salmon, rare wildflowers, or certain varieties of crayfish or butterflies. It is often assumed that such circumstances existed in earlier times in which targeted species are further assumed to have existed in greater numbers.

These theoretical constructs have then been integrated into various laws and regulations in which management actions are focused on "restoring" such perceived conditions under the supposition that target species will also be "restored" to past presumed numbers.

A principal problem with these strategies is that the existence of people is typically not factored into the theoretical descriptions of the past. It is also apparently assumed that the past was more stable and less dynamic than the present and that extended droughts, volcanic eruptions, major wildfires and other events were little more than "perturbations" to the established order of the environment, which then must have necessarily returned to the imagined equilibrium. Documentary research and field observations typically disprove these theories, but by then it is often too late to reconsider the rules and regulations that have been established in response to these misconceptions.

Figures 4 and 5 show historical examples of landslides, wildfires and floods that have converted forested habitats to fish habitats within a matter of minutes and days. These abrupt transitions resulted in immediate changes in local plant and animal populations and are typical of the types of dynamic events that define forest history throughout the Pacific Northwest.

The Fire Triangle

The most consistent and widespread form of disturbance in our region's forests is people, and for the past 15,000 years their principal management tool and agent

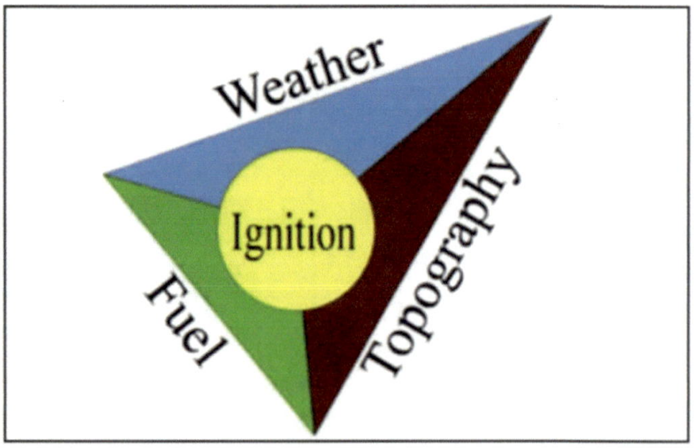

Fig. 6. Fire Triangle. Some versions of this model use oxygen and heat instead of weather and topography.

Fig. 7. Mt. St. Helens eruption of 1847 resulting in major forest fires.

of change has been fire. People are the only animals who can start fire and that advantage has been used on a daily basis since they first began to occupy the Pacific Northwest.

Figure 6, the "fire triangle," illustrates how fire is constrained across the landscape. First, fire must be created: this occurs occasionally along the Cascade Range by volcanic eruptions (Figure 7); seasonally in some areas of the region via lightning strikes (Figure 8); and on a daily basis by people everywhere (Figure 9). The duration and extent of each ignition is dictated by available fuel, local weather conditions, and – if the fire begins to spread away from its point of origin – by topography.

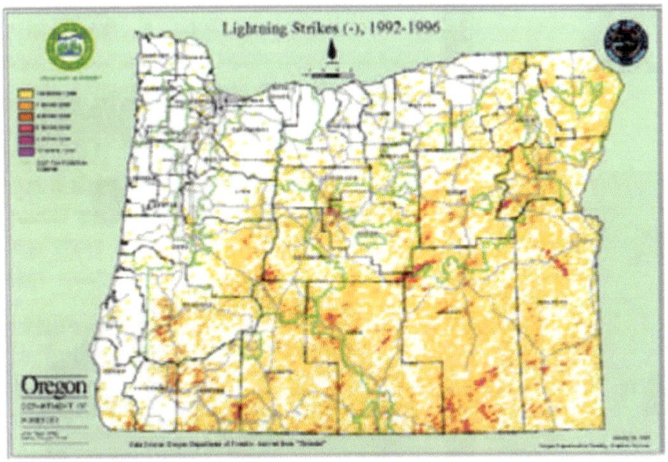

Fig. 8. Typical Oregon lightning strike patterns for the years 1992-1996.

Fig. 9. Backfire ignited on Warm Spring Reservation in 2012.

It is generally recognized that forest fires in the Pacific Northwest tend to take place in late summer droughts when fuels are dry and humidity is low, that these conditions are exemplified during east wind events, and that fire travels uphill better.

The transformative effects of fire are also widely recognized, where in a matter of minutes, hours, and days entire creek basins are changed from conditions of protective shade and vegetative cover to barren landscapes of smoking snags, charred earth, and polluted air and waters (Fig. 10). Plants and animals that are not killed outright during these events find that their habitats have changed dramatically and irrevocably (Fig. 11). Many plants that survive these events find competition for sun, seedbeds, and other opportunities to grow and reproduce are greatly enhanced; whereas most animals are forced to relocate to undamaged areas in order to obtain food and shelter in order to continue living; greatly favoring surviving predators and putting plant eating species at ever greater risk.

Conclusions

When I attended forestry classes at Oregon State University in the late 1980s and 1990s there was a lot of interest in such concepts as "preserving old-growth forests," "maintaining spotted owl habitat," and "riparian enhancement." These concepts were typically rationalized by unfounded theories of "steady state ecosystems" and

Fig. 10. 1933 Tillamook Fire: 280,000 acres burned in 24 hours with 8 mile high mushroom clouds.

idealistic descriptions of such circumstances as "non-declining, even-flow, naturally functioning" forests and grasslands. It didn't seem to matter that such conditions had never actually been observed in nature, measured, or documented – only that, for some reason for some people, they were desired.

Many of the current laws and policies governing our federal, state, tribal, private and municipal lands are based upon these theoretical misconceptions. The fact that they are inherently racist and strongly biased against past cultures and current populations has gone largely unnoticed and unchallenged. Why that is might be hard to explain, but the fact is that these biases are based on personal values and political realities far more than scientific findings or research. The additional fact that these laws and regulations are claimed to be science-based is even more troubling.

The practice of science has been seriously compromised during this process, and our rural economies and environments have been signficantly damaged as a result. The same argument can be made regarding related damages to our native plant and animal populations, the degraded quality of our air and waters, as well as the teaching of science itself.

From my perspective, it is time to again return to basic scientific principles of observation, documentation, experimentation and replication in the management of our natural and cultural resources. The recent fad of producing complex computerized "models" of our circumstances and desired future conditions was made possible by incredible recent advancements of technology that permitted sophisticated representations of actual environmental conditions. Graphic simulations and mathematical interpretations have been used to replace actual reality; traditional field scientists have been replaced by computer gamers; and voters and politicians have somehow become convinced that these are inherent improvements in our abilities to understand and manage our physical environments and the resources they contain.

After more than 30 years of this charade it has become obvious that digitizing reality and expecting computers to foretell our common futures has not worked as promoted.

There is a place for computerized modeling and the predictive capabilities of advanced mathematic simulations, but that place is not a replacement for actual reality or for human decision-making capabilities.

It is time we returned to traditional scientific methods and put our faith in legitimate experience and earned knowledge if we are to leave our descendants something similar to the wonderful conditions we were given by our own ancestors on the land. In my opinion.

Fig. 11. Second-growth Tillamook snags resulting from 1933 fire.

Forests And Our History ... Or, How Did Things Get This Way?

By Dick Powell, Retired Western Oregon Forester

In mid-January 2015, the Oregon Society of American Foresters in partnership with the Oregon Forest Resources Institute, hosted the annual Forestry Leadership Academy. This was for forestry leaders from both Oregon and Washington that included the Society of American Foresters, small woodland owner's association, Tree Farm, and others from the two states. I made a presentation that dealt with the relationship between the development of human civilization and forests. It was well received and it was suggested the gist of the presentation be written as an article.

To effectively communicate about forests to the public, a forestry leader needs to understand the history of the forest and the role forests played in human civilization – both past and present. Schools are good about teaching history and biology as distinctly separate subjects but are very poor about teaching history AND biology as a single subject. In other words, when looking at the landscape, one is looking at both its biology AND its history. History and biology are so intertwined they cannot be separated if one is to understand the landscape.

WOOD FOR FUEL, MANUFACTURING AND TRANSPORTATION

At the beginnings of mankind, once people learned that fire was a useful tool and could be made on demand, people also understood that a source of fuel was necessary. A readily available fuel was wood.

Then, as civilizations evolved from hunter/gatherers and became more agrarian with stable population centers, they found that wood was useful as a building material as well as a source of fuel.

At some point, people discovered that clay could be fired and made into pottery amphora and pots. These allowed people to store grains and wines for future use and more easily transport their goods. Firing clay required heat and wood was a very available fuel.

Later, it was discovered that copper and other metals could be extracted from the earth and shaped into useful objects. This led to the Bronze Age and a huge step forward in human civilization. Again, wood made the smelting and shaping of metal possible.

Initially, people moved about and transported their goods by foot. Then, someone probably noticed that wood floats and could be shaped into rafts and ships. These greatly facilitated the movement of people and eventually large quantities of goods across the seas and to distant lands.

When the wheel was invented and carts and wagons came into use, it became much easier to transport larger quantities of goods and for people to travel greater distances. Wood was light, easily shaped, and available.

FORESTS AND EROSION

Though they do have forests and foresters (including Saudi Arabia), most of us think of the Middle East as a dry, desolate, barren, and treeless land. However, at one time, it was once a land of forests. The problem was that the idea of sustainability and reforestation was unknown to those early civilizations. When the hillsides were stripped of their trees and heavily grazed by sheep and goats, soil tended to erode.

Ancient Greek Pottery.

Mesopotamian Clay Jar, 5000BC

The Roman city of Ephesus on the west coast of today's Turkey was thought to be the third largest Roman city in Asia Minor (around 50,000 people) and a bustling seaport that opened onto the Aegean Sea. Due to deforestation, over-grazing, and erosion and despite attempts to divert the river and dredge the harbor, Ephesus' harbor silted in; today, it is five-six miles inland!

When a civilization ran out of wood they sometimes looked to their neighbors. The Romans, for instance, had so deforested the Italian peninsula that they had trouble firing their baths. Wood was needed to produce silver coins and the shortage of wood was affecting their monetary system.

Bronze Age Artifacts.

Phoenician Ship, 700 BC

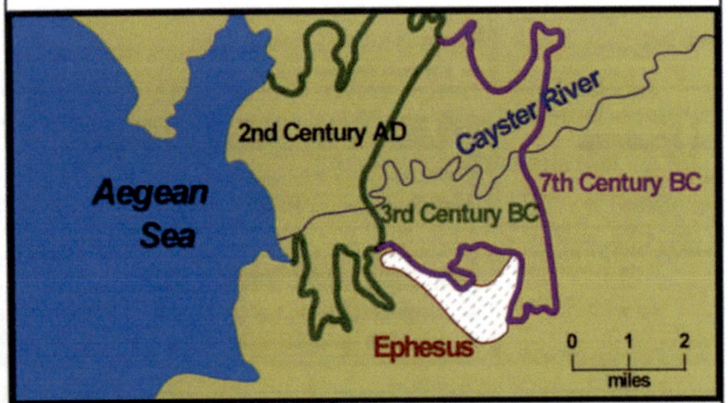

Ephesus - The Adriatic coastline as it was in the 7th Century BC, 3rd Century BC and 2nd Century AD

Furthermore, they were having trouble building the wooden ships necessary for the maintenance of the Empire. In part, the Romans conquered their neighbors for new sources wood.

FORESTS, HUMAN DISEASES AND FIRE

Here in the New World (both N. and S. America), the indigenous people were sophisticated managers of their land. Because there were no easily domesticated large herbivores (elk, bison, etc.), these people did not have livestock as was found in other parts of the world and they had no need for pastures or fences. Instead, they manipulated the landscape to provide better forage for the game they hunted and for the crops they grew. [One might say these were their "pastures" but, without fences and plowed fields, these were not recognized for what they were by the Europeans.] Their favorite tool was fire.

When Hernando de Soto traveled through the southeastern US in 1540, he found large, stable, agrarian societies and sophisticated population centers. He found fields of four square miles! Clearly, the people he encountered were not nomadic hunter/gatherers.

But, his travels and his animals spread Old World diseases. These diseases (cow pox, chicken pox, swine flu, etc.) were unknown by the Indian because they did not have livestock they lived. They had built up no immunity to deadly diseases nor did they have medicines or other means of coping. Consequently, it is estimated that 90-95% of the New World's native population died from these introduced diseases.

When de la Salle travelled through the southeastern US 140 years later, the civilizations found by de Soto had all but disappeared.

When so many of these people disappeared from the landscape, their management practices also ceased. Though native, trees can be very invasive and they soon occupied the previously Indian-managed landscape.

FORESTED LANDSCAPES AND WESTERN CIVILIZATION

By the mid to late 1700's, 250-300 years had passed – plenty of time for trees to invade large areas and become well established forests. By now, the Euro-Americans had come to dominate the landscape in colonial America and they were pushing over the Appalachians and westward toward the Mississippi River. They encountered this forest and thought it a wilderness and an obstacle to settlement. This "wilderness" entered into American lore and texts during the Romantic Period of American history (early 1800's) though it might more correctly be described as an artifact of introduced Old World disease!

The settlers cleared land for their farms and for fuel, fences, and building materials for their homes and barns and their new cities. As they advanced westward, they began to clear that forest.

With the advent of the railroad in the mid-1800's, large numbers of people could travel easily and ship their commodities over long distances. Other than the locomotive, rails, spikes, and the car's undercarriages, the train was a wooden enterprise as the cars and railroad ties were wood. Very importantly, the fuel was wood. By the late 1800's the railroads were using 15-20 million acres of forest a year! (It is probably fair to say that the wooden railroad hugely influenced the settlement of the American West.)

WOOD PRODUCTS AND MODERN TECHNOLOGY

By 1920, American energy needs from wood decreased from about 90% in 1850 to about 10%. Up until 1900, virtually all iron was produced with charcoal (a wood product).

An image of the American "wilderness" as depicted during the Romantic Period of American history (early 1800's) that entered into American lore and texts. Today, this image is probably very much what people have when they think of the "wilderness" encountered by the settlers.

By the late 1800's and early 1900's, things were changing. Coal and oil had been discovered and were coming into widespread use as an energy source.

People had discovered wood preservatives and railroad ties were lasting 30- 40 years instead of 5-7 years. Farmers had improved crop varieties and developed pesticides. Machines were replacing draft animals on the farm and in the city.

As a result, farms were becoming much more productive and less farm land was needed to feed the nation; in other words, the excess farmland was being abandoned. Being rather invasive, trees soon occupied this land and became much of the forest we know today.

If the reader is trying to keep score, it seems that, as the American Indian was taken out of the landscape, the number of forested acres in the US increased for a time. During the westward movement of the Euro-American beginning in the mid to late 1700's, the forested acreage decreased until coal and petroleum began replacing wood as an energy source in the late 1800's. At that point, the acres of forest land began to increase. Since the early 1900's, the forested acreage in the US has stayed fairly constant.

FOREST HISTORY AND THE FUTURE

Understanding the development of the forested landscape as it relates to human history helps people understand the forests we see today. For example, today, worldwide, a little over half of all wood use is for heating and cooking. Thousands of things we use on a daily basis come from trees. With an understanding of yesterday and looking at today and tomorrow, the real questions are: what do we want from our forests? .... what do we want for our forests? And, very importantly, how do we get there?

Ephesusas it looks today. My guess is that the flat, marshy ground below the Roman theater is the silt that eroded off the hills and filled in the harbor. The Arcadian Way (2000 ft. long and 35 ft. wide) was built by Emperor Arcadius in the 1st Century BC, probably to maintain access to the sea. In the distance and between the hills seems to be the Adriatic Sea as it is today.

The Caddoans as de Soto found them (in the Arkansas/Texas region). Note their pyramids and ceremonial mounds in the background. Hardly what one would find with a nomadic hunter/gatherer society.

Forest Restoration:
Problems and Opportunities Revisited

By Dr. Bob Zybach

How actively managing our western forestlands on a landscape-scale can immediately create thousands of rural jobs, greatly reduce catastrophic wildfire risks and damages, return millions of dollars to our state and federal treasuries, increase native wildlife populations, fund our rural schools, roads, and libraries, and make our forests and grasslands safer and more beautiful than ever before. Seriously.

Photograph 1: Oak Type. Former oak and pine savanna.

This article is an updated version of the first article I wrote for this magazine, in the Spring 2012 issue. Since then we have continued to witness catastrophic wildfires in northern California, Oregon, Idaho, and Washington that have burned millions of acres, killed millions of wildlife, grossly polluted the air in our major towns and cities for weeks on end, and killed dozens of people. All predictable, and all preventable if corrective actions, generating thousands of jobs and millions of dollars in profits and tax revenues, had been taken first. What was true seven years ago remains true today; and for tomorrow, too.

Western forestlands have never been in worse shape: millions of acres of dead and rotting trees; thousands of miles of abandoned and barely maintained roads; record setting wildfires becoming larger, deadlier, and more destructive by the year; hundreds of artificially impoverished rural communities; and endless litigation preventing the use of resources we need to sustain our lives and our economy.

There are a number of reasonable ways to resolve these problems; a long-term commitment to active forest restoration and management would seem to offer the most immediate benefits to both people and wildlife, and to be the most likely route to long-term economic sustainability as well.

What is forest restoration, why is it needed, and how is it done are the questions addressed in this article. Two examples of current forest restoration projects are profiled to help answer these questions, and to illustrate how these types of programs can be immediately implemented across the landscape to the benefit of neglected forests and depressed timber-dependent communities throughout the West.

What is Forest Restoration?

The process of forest restoration is focused on returning an area to one reflecting desired past conditions. It is critical to understand a) what conditions were actually like in the past, and b) which of those characteristics (if any) should be restored or preserved for the future.

For the past 10,000 years and longer, people living in Oregon have used and managed native plants and animals for their own purposes: principally for food; shelter; fuel; and fiber products, such as clothing, basketry, musical instruments, canoes, ropes, and weapons. Fire was used for a wide range of purposes: for cooking, heating, and lighting areas around homes and campgrounds; for rejuvenating berry patches and harvesting fields of grain; for hunting game by systematically setting vast tracts of land on fire.

Man is the only animal that can use fire, but he is not the only animal that benefits from it. The expert and judicious use of fire across the ancient landscapes of Oregon resulted in the stable patterns of forests, woodlands, vast prairies, wetland meadows, brakes, balds and berry patches encountered by Oregon Trail immigrants when they first arrived here in the 1840-1850s. The great numbers of elk, deer, birds, fish, squirrels, migratory fowl, and other animals that populated these environments were noted and documented by many of the new residents.

By that time, though, most of the Indian communities in this region had been decimated by plagues of deadly diseases introduced by Asian, African, European, and American explorers, hunters, and traders in earlier decades. The new arrivals, encountering a land nearly devoid of native people, assumed they were "settling" a "natural" landscape created by their God or by Nature, just for them. "Manifest Destiny." That mistaken belief persists to this day in our popular books and films, in our scientific literature, and in our legal system, and is a key reason so many of our forests and grasslands are in such degraded conditions at this time.

Photograph 2: Pine Type. Invasive Doug-fir/madrone.

Photograph 3: Doug-fir Type. Douglas-fir/pine mix.

Forest restoration, more than any other definition, means restoring people to the land; and restoring them in such a way that they feel safe, whether in the woods, along a river, or walking through a town. Restoring people to the land also supposes restoring fire to the land; fires set by people, not by lightning.

Upper South Umpqua Project: Considering Past Conditions is Step 1.

The map shown in this article represents a critical step in the forest restoration process – a determination and documentation of likely past conditions for areas being considered for restoration. This step is quite often ignored, or even unrecognized, invariably resulting in failed projects over time. Whenever we plan to restore something, it is important we understand the actual conditions – including presence and actions of people - that existed in the past.

The Upper South Umpqua Headwaters Precontact Reference Conditions Study focused on characterizing a significant portion of the Umpqua National Forest in Douglas County, as it likely existed in 1825, prior to white contact. The study area is slightly more than 230,000 acres in size and extends from the crest of the Cascade Range at elevations greater than

6,000 feet, westward to the confluence of Jackson Creek with the South Umpqua River at approximately 1,100 feet elevation. The map shows the location and composition of forest type patterns and basic travel routes as they likely existed in the study area 200 years ago. Each of the subsequent four photographs documents a typical example of the four identified forest types, and illustrates potential forest management actions needed to restore and maintain desired future conditions.

Photograph 4: True Fir Type, Huckleberry Lake.

One of the basic purposes of forest restoration is to reduce wildfire risk and damages. The method for achieving this in overstocked stands of conifers is to significantly reduce their biomass ("fuel load") and open up the tree canopies ("thinning") as they existed in earlier times, when catastrophic-scale crown fires were uncommon occurrences. On federal lands this is referred to as an "FRCC 1" condition.

The Upper South Umpqua Project was initiated by Douglas County Commissioner, Joe Laurance, to consider the possibility of restoring degraded local forestlands to this type of condition. On July 15, 2010, he testified to a Congressional subcommittee of The House Natural Resources Committee in Washington, DC:

"Fire Regime Condition Class (FRCC) 1 is similar to the forest which European explorers first found here. That forest had been modified by fire for more than six thousand years to provide the native inhabitants with what were then life's necessities. These included abundant wild game from the most productive and diverse wildlife habitat ever known on this continent. Similarly, the regular burning of competing vegetation permitted propagation of nut bearing trees and other food producing plants. Additionally, the historic "Healthy Forest" promoted pristine rivers, streams, and lakes that provided an abundant harvest of fish and waterfowl. Within FRCC 1 the risk of losing key ecosystem components to fire is low, while vegetation species composition, structure, and pattern are intact and functioning within the natural historic range."

Research methods used to determine and document 1825-era forest conditions in the study area included extensive use of General Land Office survey maps and notes, historical maps and photographs, field plots, oral history interviews, literature reviews, archival research, and over 5,000 GPS-referenced digital photographs. This latter method documented the loca-

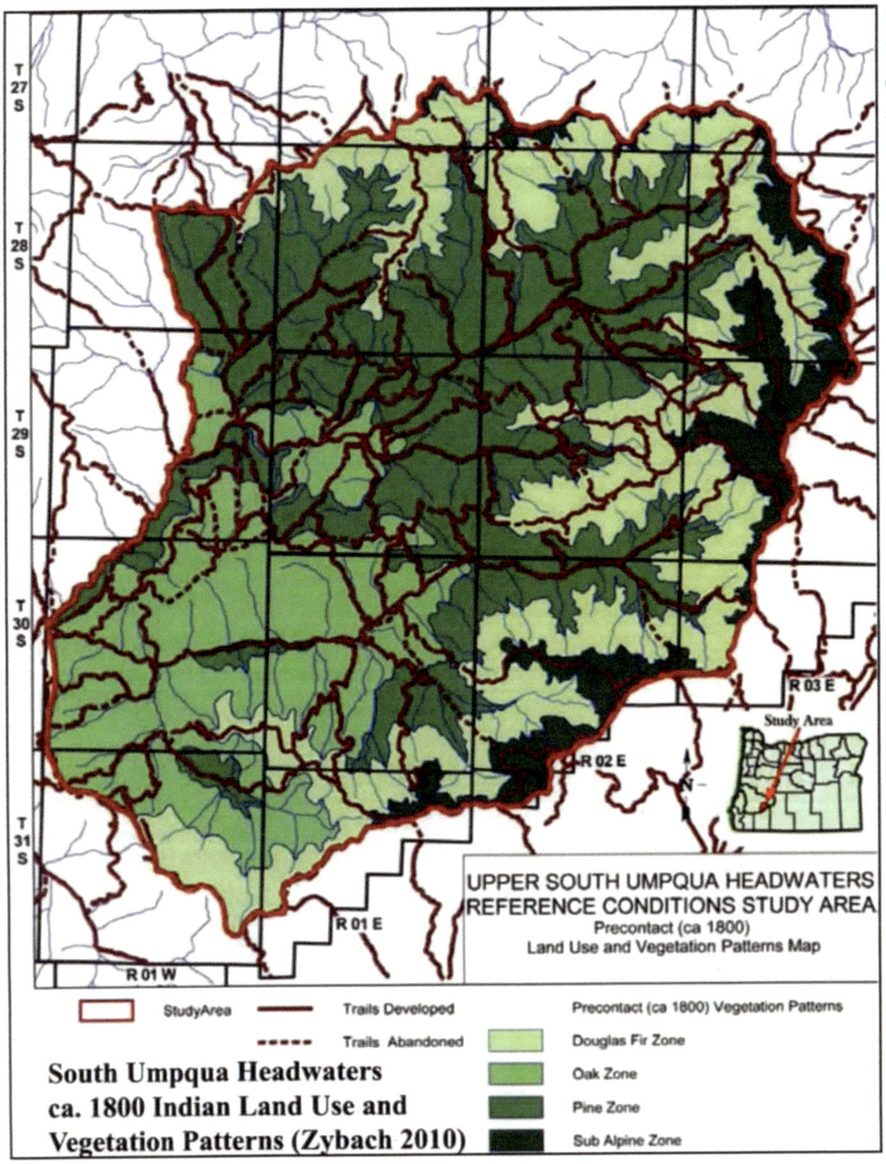

South Umpqua Headwaters ca. 1800 Indian Land Use and Vegetation Patterns (Zybach 2010)

tion and extent of remaining old-growth (pre-1825) trees in the study area, in addition to documenting persistent patterns and patches of such traditional cultural food and fiber plants as camas, fawn lilies, cat's ears, huckleberries, hazelnuts, chinquapin, tarweed, serviceberry, wokas, bracken fern, thimbleberries, and salal.

Historical research has given us the map shown: a generalized depiction of likely forest conditions in the study area during the 1800-1825 time period. The following four photographs represent typical current conditions within each of the four forest types (or "zones") that are depicted. The large size and wide spacing of the older trees in the photographs can be gauged by the "human scale" used to measure them: Nana Lapham, long-time NW Maps Co. forest science research assistant, is 5' 8" tall and did much of the field work on this project.

Photograph 1 shows relict trees from an oak and pine savanna that was developed and used by native residents for hundreds or thousands of years. These areas were tended by constant gathering of fuel, acorns, and pine nuts from the trees and regularly tilling and/or burning the surface area to manage understory crops, such as camas, tarweed, hazel, and beargrass. Notice how Douglas-fir have invaded the area in the past 100+ years, threatening the survival of the few remaining old-growth oak and pine.

Restoring this savanna would entail removing the Douglas-fir before they smother the remaining old-growth; removing the surface fuels that have built up around the bases of the oaks and pines; and reintroducing the types of understory plants and regular burning practices that created and maintained savanna conditions in the first place. There are thousands of acres similar to this throughout the study area, with invasive Douglas-fir slowly killing the established old-growth and creating potential crown-fire conditions by developing a continuous canopy of fine, pitchy fuels. A crown fire under these circumstances would likely kill all of the trees in this picture, including the old-growth.

Photograph 2 is a typical condition found throughout the former Pine Type in the study area. Again, these areas were maintained by regular prescribed fires in precontact time, and are now being threatened by thousands of invasive Douglas-fir and madrone.

Photograph 3 shows the 1825 Douglas-fir Type, which still contains scattered old-growth sugar and ponderosa pine; species which may have dominated this type in the Indian era. Restoration would provide the best option for prolong the lives of these historic trees, too.

Photograph 4 is a picture of Huckleberry Lake, in the high elevation True Fir Type. The "lake" is now a wetland prairie, a result of eutrophication during historical time. This area used to be a portion of a major huckleberry gathering complex, equally accessible to Indian families living in the Rogue River and South Umpqua basins due to its common ridgeline location, before these plants were shaded out by the invading conifers. In addition to the hundreds of acres of cultural plants still present in the area, numerous stone artifacts and other evidence of past use are also in widespread evidence along this ridgeline. The scattered taller trees in the background represent larger diameter, older trees likely dating to the 1600s and 1700s; the vast majority of trees, though, having established themselves after the mid-1800s and throughout the 1900s.

The common theme documented by this map and photographs is that tens of thousands of acres of old-growth oak, pine, Douglas-fir, red cedar, and other conifers exist throughout the study area in need of immediate attention, if they are

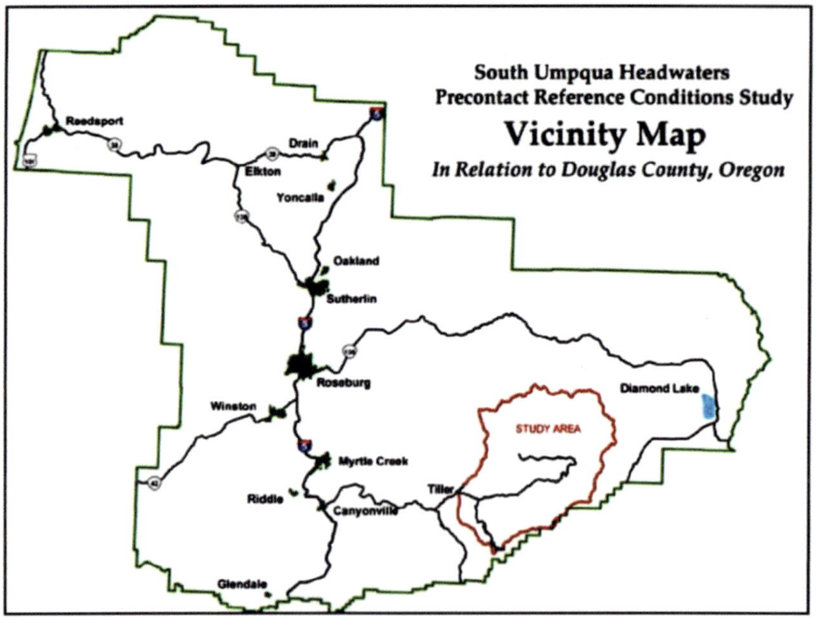

to survive [Note: 17,000 acres of this study area burned in the 2013 "Whiskey Fire," one year after this was written, costing $23 million and needing 932 firefighters to contain]. The same problem exists for the scattered patches of huckleberry, camas, tarweed, and native grasses that still persist. A forest is, ideally, composed of many facets, housing many different types and species of plants and animals. Those are the types of attributes that used to define these lands, and the same types that can be restored and maintained for future generations of people and wildlife.

Jims Creek Project: Make a Choice and Then Do It is Step 2.

The second step to forest restoration, after considering historical conditions ("options") is to determine what future conditions are desired (goals and objectives), and to begin actively restoring and/or maintaining those prized and desired conditions across the land. An excellent example of this step is the Jims Creek forest restoration project on the Middle Fork Willamette Ranger District which, in 2012, was in the process of completing an initial 400-acre "demonstration project" portion of a 25,000-acre plan.

The Jims Creek project has demonstrated the feasibility, profitability, and general benefits of conducting landscape-scale forest restoration projects on federal forestlands, but it is also a good example of how much time and money can be spent in putting these projects together, as well as the ease and quickness with which they can be stopped by adversarial legal actions. This project was initially conceived by a local US Forest Service forester/project manager, Tim Bailey, who then spent the better portion of the next ten years shepherding his vision through the myriad public meetings, scientific reviews, committee presentations, promotional tours, and other hurdles needed to get things underway on the ground.

Picture 5 shows Bailey in front of a portion of the Jims Creek Project in 2010. This area had already been treated by removing most of the invasive conifers established during the past century, and by broadcast burning the ground so as to remove excess litter and logging debris. Note the scattered trees that have been left behind: widely spaced pine of several different age groups, from seedlings to saplings and second-growth to old-growth. Also note the small herd of elk grazing directly above Tim, near the crest of the hill.

Picture 6 is the same herd of elk as the previous picture, seen through a zoom lens. Not shown in this photograph are several more elk just over the crest of the hill and at least two blacktail deer near timberline that were photographed running deeper into the woods shortly after this picture was taken. The small charred stumps and large woody debris in the foreground will soon rot away or be consumed in the next few surface fires. The reddish-brown pattern is bracken fern, a plant harvested in large quantities by many Oregon tribes for its starchy roots and asparagus-like "fiddleheads" that grow in the spring. In a few more years, with a few more broadcast burns, this area will appear very similar to what it must have looked like hundreds of years ago -- including the regular presence of elk and deer.

Although the 400-acre Jims Creek demonstration project has clearly shown several advantages of forest restoration in this area, additional progress has been halted at this time due to an infestation of thousands of red tree voles ("tree mice") that accompanied the migration of Douglas-fir trees into the project area during the past century. These rodents are protected against logging Douglas-firs under a federal "survey and manage" regulation, despite the fact they are not a threatened or endangered species.

This type of work stoppage, based on relatively new federal regulations and related litigation initiated by environmental organizations, has become the main difficulty in beginning and completing forest restoration projects in Oregon and throughout the West. The Jims Creek Project is also a good example of those types of problems: of the 25,000 total acres of this project within the Middle Fork Ranger District, 7,000 acres are privately owned and being managed for maximum timber production; more than 7,000 acres have been classified as spotted owl habitat; approximately 7,000 acres are in remote areas that would likely include the "taking" of spotted owls; and the remaining 4,000 acres are populated with regulated tree voles. Also, a river flows through the project area that contains two listed fish species and "there is a perception on the part of the regulatory agencies that this type of restoration work can have a negative effect on fish."

Despite these hurdles, there is a lot of work needed and a lot of people wanting to do it.

Conclusions

Forest restoration projects should be conducted on a landscape-scale basis in order to be effective biologically, aesthetically, and economically. Project boundaries should include sufficient commercial materials to treat the entire project area and to show a profit. Profitable and beneficial actions

Photograph 5: Tim Bailey and elk herd on Jims Creek site.

are sustainable on a long-term basis, as we have learned from more than 10,000 years of forest history in this region.

The actions needed to restore our forests to earlier, more desirable conditions would necessarily create thousands of jobs for decades -- jobs to make the best uses of our common resources, protect our old-growth and wildlife, and greatly reduce the likelihood of wildfire and the severity of such occurrences when they do take place.

Based on my own experiences and observations, I think there are four key things that must be in place for forest restoration projects to be successful on a long-term basis:

1) Areas slated for restoration should include sufficiently broad boundaries and specifications to allow projects to be profitable;

2) Restoration projects should be landscape-scale (25,000 to 250,000 acres) in size in order to be economically efficient and biologically effective over time;

3) Local residents and businesses should be in strong support of restoration projects, and be given access to all information that develops during the process;

4) Local project managers should be knowledgeable and capable of communicating scientific, technical, and political aspects of a project to local citizens.

I remain certain that the adoption of these practices, as defined, would have many immediate and positive effects on forest health, old-growth preservation, endangered species protection, rural economies, international trade balances, and many other economical, ecological, cultural, historical, aesthetic and recreational values associated with Oregon's forests.

The degradation, destruction, and loss of our federal forests and grasslands to wildfire, bugs, and disease will continue to escalate so long as we continue our current path of passive avoidance and neglect. Restoring our Nation's forests means restoring people – in part, as active managers – to our lands. The benefits for doing so have been listed; the impediments to getting started have been largely self-inflicted, are almost entirely political (rather than scientific or humanitarian), and can be readily surmounted, given effective leadership of common outcry. The best time for doing something is now.

Photograph 6: Close-up of 15-16 elk in restored prairie.

The Osbornes: Historical Documentation versus "Critical Habitat"

By Dr. Bob Zybach

Crater Lake and Wizard Island as viewed from The Watchman "Observation Station," July 23, 1933. Photo by Lester Moe and Robert Snyder

William B. "Bush" Osborne, Jr. was one of the most important and accomplished foresters in the US Forest Service during the 20th century, and ever. Following his graduation from Yale with a Master of Forestry degree in 1909, he got his first job on the Oregon (now Mount Hood) National Forest and continued working with the Forest Service until his retirement in 1945.

In that 35-year period Osborne invented and developed some of the most important concepts, standards, and tools that continue to typify and help define the Forest Service and forest fighting to this day.

During the years 1909 to 1910, Osborne conducted an extensive reconnaissance of the Mt. Hood and Santiam National Forests, followed by his drafting of the first detailed maps of a national forest. The year 1910 is well-known for the widespread catastrophic wildfires of that year. Osborne helped fight a number of these, and his familiarity with the Mt. Hood and Santiam peaks, forests, and vistas led him to strategically plan and install the first coordinated lookout stations in the Pacific Northwest.

The intended purpose of Osborne's maps was the location of wildfires from these viewpoints, but they were not accurate enough for his intentions. In 1911 and 1912 he did the "surveying, compilation, and drafting of the first trigonometrically controlled national forest map." This methodology and attention to detail soon became the standard practice for all national forests. It also typified Osborne's career-long insistence on "pin point accuracy."

To further increase the ability of seasonal lookouts to accurately locate fires, Osborne invented the "Oregon fire finder" in 1913: a precise instrument that worked in coordination with his maps. He made improvements to his invention in 1916 that became the standard for forest lookouts and was eventually known as the "Osborne Firefinder." By the time of his retirement in 1945, over 4,000 Firefinders were still in active use. Today these devices are collector's items, but modern versions continue to be used to this time as originally designed.

Less well-known than his Firefinder was Osborne's invention of the "photo recording transit" in the late 1920s. By 1933 his specially-designed and thoroughly tested camera/transit was ready for use, and during the next 10 years more than 1500 lookout stations in Oregon and Washington were using them.

The following pages contain several examples of Osborne photographs taken from western Oregon fire lookouts in the 1930s. The original prints are much larger -- about 14-inches by 4-inches -- and all were taken in groups of three. These photos covered the entire 360-degrees of visible landscape, beginning with a due-north photo and proceeding southward, east to west, at 120-degree increments. The specially-designed camera used a curved negative surface so that there was no distortion on the outer edges of the photographs. These products can thus be described, without exaggeration, as highly detailed and accurate maps of specific forest vistas at particular points in time.

1936-1937 Lincoln County "Critical Habitat"

Lincoln County is the birthplace of identifying marbled murrelets as an endangered Oregon species, beginning with the discovery of a fledgling in a Siletz grocery store parking lot in 1977. Subsequent research resulted in discovery of a nest with a single egg in a "300-year-old" Douglas fir tree, leading to immediate speculation that these birds required "old-growth" conifer trees for their survival.

Most of the forested lands and prairies of Lincoln County burned in the 1849 Yaquina Fire, and burned again in 1868. The "old-growth" Douglas fir turned out to be a relatively-young, though large, 100-year-old second-growth. The vast majority of these birds live in Alaska and do not nest in trees. They can fly 60 to 90 miles an hour and spend their entire life – except nesting for one month a year some years – in the ocean eating fish. The idea they require big, old conifer trees to exist is nonsense. There is a big difference between "preferred" habitat and "critical" habitat in the dictionary.

In 1992 the "big lie" was promoted by the federal government that nearly 90% of the Oregon Coast Range had been clearcut, beginning in 1850. The imaginary widespread logging had supposedly resulted in a 60% to 80% reduction in the Oregon marbled murrelet population, with no information to support that assertion. Because no such information existed.

This fabrication, repeated at the highest levels of the federal bureaucracy, resulted in a 1992 "finding" that the so-called Oregon population of this bird was "threatened or endangered" because of logging. This manufactured story was largely responsible for the listing of the bird and has resulted in a lucrative cottage industry for a handful of lawyers, college professors, and agency employees. The cost to taxpayers has been billions of dollars, the loss of tens of thousands of blue-collar jobs, and the degradation of dozens or hundreds of timber-producing rural communities.

Despite the easily-disproven nature of these assertions, they have continued to be repeated as if fact for decades; the most recent instance being a Portland Audubon website focused on "Oregon's" marbled murrelets two years ago.

Marbled murrelets are not dependent on "old-growth" – or even big conifer trees – for their survival. The statements regarding logging history of the Coast Range are not true and easily disproven. There is no information – one way or another – regarding populations of Oregon marbled murrelets prior to 1992. None. They may have even migrated here, much like barred hoot owls in recent years; possibly following the reestablishment of coastal forests following the "Great Fires" of 1840 through 1951. Or maybe not.

Most marbled murrelets are based in Alaska and there is no substantive evidence that logging (or burning) coastal forestlands has had any effect on their existence.

The Lincoln County map of historic forest fire lookouts uses a 1951 "Metsker" county index map showing the page numbers of the larger-scaled maps included in the atlas. Typically, individual Metsker maps are made up of a single square 23,000-acre "township" of 36 square-miles. These maps are highly detailed -- including most landmark names, major landowners, roads, trails, and structures within the township -- and are excellent resources to help interpret historical Osborne and aerial photos for similar time periods.

Map of Lincoln County 1930's Osborne Forest Fire Lookouts (red), 1940's WWII Lookouts (black) and representative logging operation fire lookouts (blue) on base 1951 Metsker Map.

South Beaver Lookout

At least six sets of Osborne photographs were taken in Lincoln County from 1934 to 1937. The South Beaver Lookout was the westernmost location of these photos and Chinquapin Point Lookout was the easternmost. Between these two landscapes, and extending eastward to Benton County, is the greatest known and most studied concentration of breeding marbled murrelets in Oregon.

The May 29, 1937 northern view from South Beaver is bounded by the Pacific Ocean on the northwest and Alsea Bay to the northeast. The landscape is typified by a few young Douglas fir trees in the foreground and a likely plantation of similar-age Douglas fir in the middle distance, as indicated by the linear boundaries and apparent uniformity in size and spacing. No old-growth trees or snags are apparent. Note the road and fenceline for scale.

To the east, looking toward Table Mountain, is a sea of alder with a patch of Douglas fir in the distance to the southeast, and scattered second-growth Douglas fir beginning to emerge through the alder canopy. A few scattered snags and older-

Northern View from South Beaver Lookout, with Pacific Ocean and Alsea Bay along the horizon.

Eastern View from South Beaver Lookout, with Alsea River Canyon and Table and Burnt Top Mountain.

Western View from South Beaver Lookout, with Pacific Ocean and Highway 101 on horizon.

Northern View from Chinquapin Lookout, Showing Young Douglas Fir Trees and Open Meadows.

Eastern View from Chinquapin Lookout, with Marys Peak, Indian Prairies, Fencing and Pruning.

Western View from Chinquapin Lookout, with Table Mountain, Meadows and Scattered Older Trees.

appearing fir can be spotted. To the west, extending to the ocean and Highway 101 is a similar landscape: a vista of flowing alder with young fir emerging above the canopy here and there. No apparent "critical habitat" for marbled murrelets.

Chinquapin Point Lookout

The September 16, 1936 view north from the Chinquapin Point Lookout was dominated by young Douglas fir trees and evidence of older prairies and meadows. This pattern is more pronounced looking eastward toward Marys Peak. Here we see the remnants of old Indian prairies, often dominated by edible brackenfern.

"These prairies were subsequently maintained by white settlers, who continued to seasonally burn them in order to regularly restore high-grade grazing areas for horses, cattle, and the local mohair goat industry." Rural western Oregon settlers who maintained these old prairies and meadows in the style of their Indian forebears, were called "fern burners." Osborne considered this practice as "the greatest enemy of reforestation in the Douglas fir region."

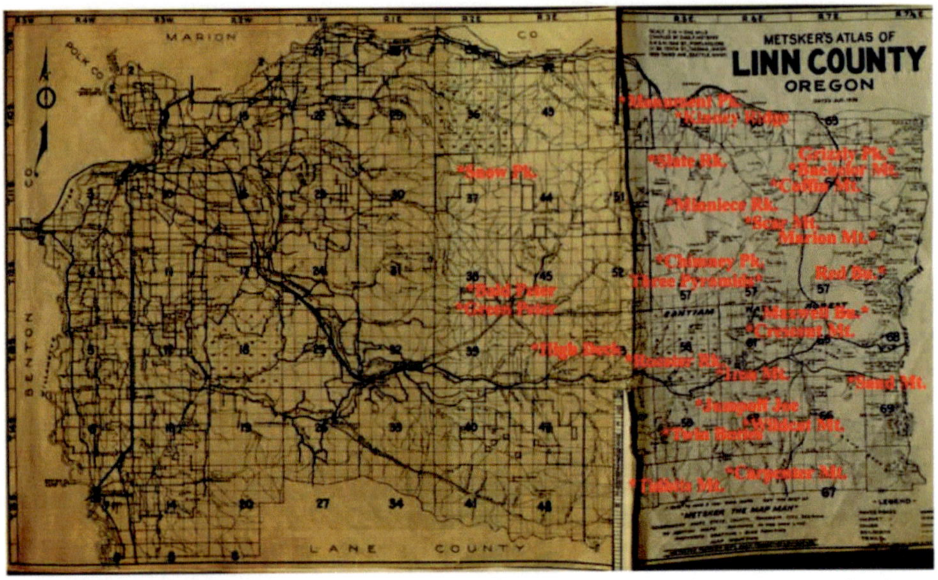

Map of Linn County 1930's Osborne Forest Fire Lookouts showing concentration in Willamette National Forest in Western Cascades; None in Eastern Willamette Valley Farmlands, on Base 1938 Metsker Map.

A major function of early lookouts was spotting and stopping fern burners. A Chinquapin Point lookout once spotted a fern fire by one of his neighbors and went to put it out. When he returned, the tires on his Forest Service vehicle had been flattened. Keep Oregon Green was created a few years later with "Don't Burn Fern" as their slogan.

The westward view from Chinquapin Point also consisted of young fir and scattered meadows, but with occasional snags and a few older, larger trees in evidence. This is the heart (western slopes and foothills of Lincoln County to Marys Peak) of so-called "critical habitat" for marbled murrelets today.

1937-1981 Linn County "Critical Habitat"

Although marbled murrelets have likely visited and flown through Linn County, there is no record of their nesting there. Rather, the "critical habitat" regulations that resulted in greatly reduced logging and reforestation schedules for these forests was related to spotted owls.

To "maintain the population" of these birds, a series of perfect cookie-cutter circles were drawn across the map -- with a focus on enclosing older and larger trees and a specific stated intent of stopping active management in those areas. There is no scientific or otherwise documented evidence that this strategy – which has also cost Oregonians billions of dollars, the loss of thousands of family-support, blue-collar jobs, and dozens of economically damaged rural communities – has resulted in the sparing of a single owl.

Instead, the federal government is spending tens of millions of taxpayer dollars to have employees hunt barred hoot owls because they are breeding with their spotted cousins. The apparent concern seems to be they might be producing viable young. It is a racial purity problem in any instance.

As clearly predicted, the practice of eliminating active management of our public forests has resulted in ever greater and more severe forest wildfires. Black-backed woodpecker "critical habitat."

The 1938 Metsker Linn County base map shows the lack of lookouts for low-lying farmlands in western Linn County, on the eastside of the agricultural Willamette Valley, and numerous lookouts to the east, on the western forested slopes of the Cascade Mountains. This is spotted owl country. The Kinney Ridge Lookout is to the north of Linn County, overlooking the North Santiam River (currently Detroit Reservoir), and Jumpoff Joe Mountain Lookout is in southern Linn County, overlooking the South Santiam River.

Kinney Ridge Lookout

The June 4, 1937 view north from Kinney Ridge Lookout was downstream, through the burned and partially logged North Santiam River Canyon toward the present-day location of Detroit Dam. The North Santiam Highway (present-day Highway 22) can be seen in its original location before it was moved to accommodate the new reservoir in 1953. A significant number of snags can be seen on the unlogged lands, but most appear to be from second-growth trees.

The eastern view from Kinney Ridge shows a number of fire-killed second-growth and understory trees in the foreground. Upstream, in the distance, is the original site of the town of Detroit. The canyon is now a major reservoir and Detroit has been moved to a different location. The western vista also shows a number of small, densely distributed snags with an emerging understory of young Douglas fir.

Wildlife populations have had to adapt to these greatly changed conditions for many decades, as is typical throughout most of western Oregon and the Pacific Northwest.

Jumpoff Joe Mountain Lookout

The July 16, 1937 view east from Jumpoff Joe Mountain Lookout showed an extensive logging operation coupled with a large burn characterized by relatively small-diameter snags and emerging Douglas fir saplings. This is the Sevenmile Creek drainage with significant stands of older trees sprinkled

Northern View from Kinney Ridge Lookout, with Burned and Logged North Santiam River Canyon.

Eastern View from Kinney Ridge Lookout, with North Santiam River Canyon, Detroit and Burned Timber.

Western View from Kinney Ridge Lookout, with Burned 2nd Growth and Conifer Saplings.

1933 Eastern View from Jumpoff Joe Mountain toward Squaw ("Latiwi") Mountain in Southeast

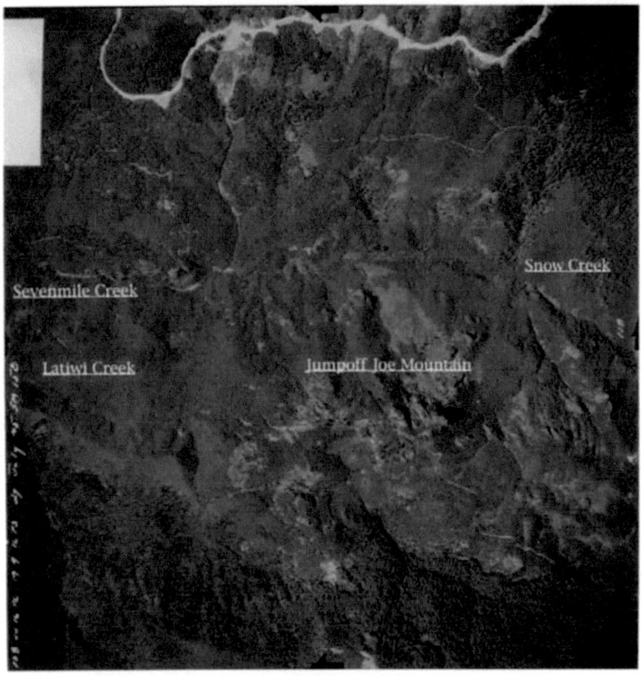

1944 Aerial Photo of Jumpoff Joe Mountain.

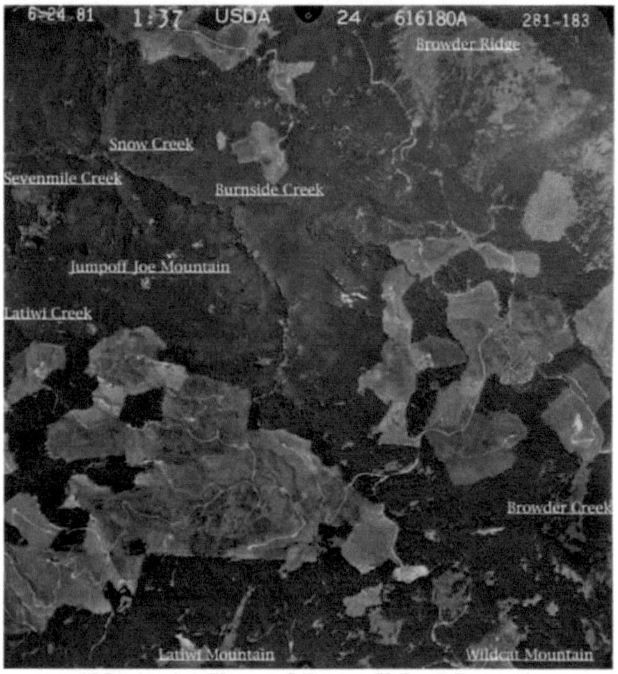

1981 Aerial Photo of Jumpoff Joe Mountain.

with snags to the east on Browder Ridge and to the southeast on Latiwi Mountain.

The 1944 aerial photo of Jumpoff Joe Mountain clearly shows this same Sevenmile Creek clearing. By 1981 the basin had fully reforested, but significant clearcut logging had extended outward on Browder Ridge and south toward Latiwi Mountain.

In the 1890s this was considered one of the best deer and elk hunting areas by Indians from Warm Spring Reservation, who hunted here by horseback. Major huckleberry fields, thimbleberry patches, camas meadows, and beargrass prairies in the immediate neighborhood indicate that this had likely been a major gathering and harvesting area for regional Indian families for hundreds or thousands of years before the arrival of horses.

Summary & Conclusions

W. B. "Bush" Osborne left an incredibly accurate and detailed history of our federal forests in the Pacific Northwest, and yet this information is unknown, ignored, or trivialized by federal regulators, lawyers, and university professors. Apparently because it contradicts and otherwise fails to support their lucrative taxpayer-funded computer models and legal filings.

This argument has been made in these pages many times before. Federal forest management regulations, based on computerized modeling, do not match the dynamic and well-documented histories of Oregon's forests and streams. Also, the draconian costs of clearly failed "critical habitat" designations continue to need legitimate scientific examination and economic analysis.

True science rises to the challenge. Good modelers use documentation to test their theories. There is great need for better understanding of actual forest habitat conditions and species adaptability for a wide variety of animals, including migrating fish and birds.

Reforestation of the 2020 Labor Day Fires in Western Oregon

By Bob Zybach, Ph.D

The series of catastrophic wildfires that began on Labor Day, September 8, 2020 in western Oregon was the worst in the State's history. My previous article on this topic recounted the basic statistics: at least 10 people died, nearly a million acres burned, more than 4,000 homes were destroyed, and millions of wildlife killed -- almost all within a two- or three-day period. Toxic, acrid smoke blanketed the region for more than a week.

Fig. 1, Powerline right-of-way and grazing cattle. Beachie Creek Fire aftermath. November 8, 2020 video-clip by McKenzie Peters, NW Maps Co.

The rebuilding has already begun but will take years to complete. Thousands of families have relocated, hundreds of miles of roadway have been cleared, power has been restored, salvage logging operations have been underway since the beginning, and wildlife populations are on the rebound. Spring weather and rejuvenated flowers, grasses, and shrubs are carpeting the landscape.

To restore our homes and forests in such a way as to reduce the frequency and severity of such future events should be the next order of business. Tens of thousands of "protected" acres of "critical habitat" have been destroyed and can never be replaced in our lifetimes, if ever. Timber production has been greatly affected, and particularly on private lands devoted to that purpose. How to fix?

For more than 20 years I operated a small business in western Oregon that reforested more than 85,000 acres of burned and logged lands. So far as I know, none have since been damaged by wildfire, while thousands of acres have already been logged and replanted again -- creating thousands of jobs and bringing significant incomes into our rural families, businesses, and counties. Lessons learned from those experiences can provide a helpful roadmap for moving forward.

Land Ownership & Road Access

There were five basic types of landownership affected by the Labor Day Fires: federal; state; family farm (including tree farms); industrial timberlands (whose owners also have commercial wood manufacturing facilities); and residential, both urban and rural. Each type requires access to repair and restore, and each has differing laws and regulations regarding what can, or must, be done.

Road access is the most critical consideration in the management of these properties and is also important regarding the management of future wildfires, floods, landslides, ice storms, snowstorms, windstorms and their aftermaths.

Good rock roads may be gated on industrial timberlands, while on federal lands they may be entirely abandoned or even made inaccessible on purpose. Managing the predictable large-scale wildfires on public lands is made more difficult, costly, or even impossible for this reason.

Federal lands, for example, include significant Wilderness areas that regularly burn in wildfires and in which access roads and the use of mechanical equipment is not allowed; dozens of designated Endangered Species Act "critical habitats" in which salvage logging, weeding, tree planting, or even entering may not be permitted; and hundreds of miles of fish bearing streams with their own restrictions. Affected recreational facilities and access roads have largely been closed to the public since Labor Day, with no indication as to when they might be opened again, if ever.

The Santiam State Forest bore the brunt of damage to state lands, in which more than 16,000 acres burned in the Beachie Creek Fire. This area has similar regulations as federal properties, but with a greater mandate to be actively managed for the economic benefit of local communities. It, too, has closed affected access roads and recreational facilities to the public at this time.

Family farms and residential areas are usually located along highways and county roads, and those areas were typically opened within days or weeks of being closed by the fires. Irrigated lawns and crops, including Christmas trees, filberts, grapes, and pears, showed little evidence of fire damage -- although homes and outbuildings occasionally burned within these protected environments due to wind-driven cinders or adjacent trees.

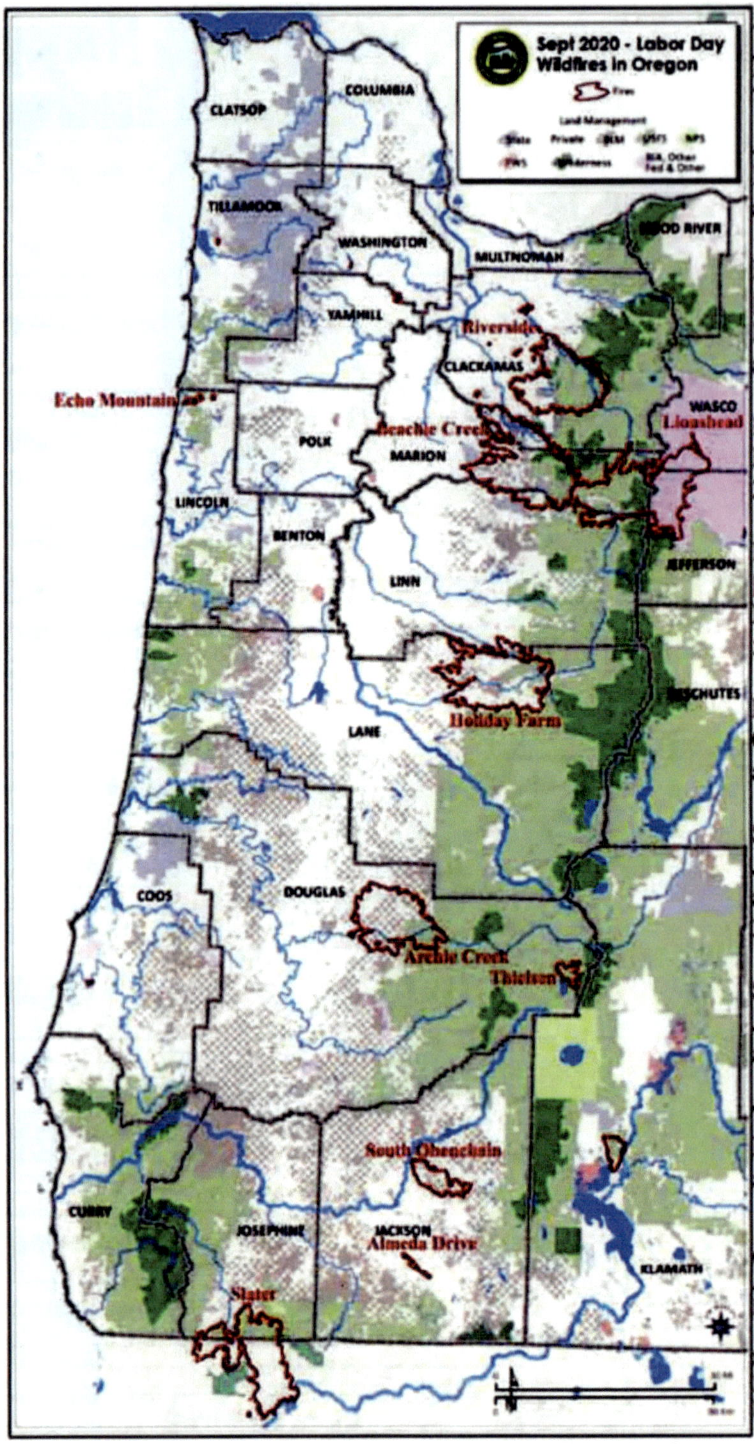

Almeda Drive.
Jackson County. 5,700 acres, 2,400 homes, 3 deaths.

Archie Creek.
Douglas County. 131,500 acres, 138 homes, 1 death.

Beachie Creek/ Lionshead [Santiam].
Marion County. 300,700 acres, 629 homes, 5 deaths.

Echo Mountain.
Lincoln County. 2,600 acres, 293 homes.

Holiday Farm [McKenzie River].
Lane County. 173,000 acres, 517 homes, 1 death.

Riverside.
Clackamas County. 138,100 acres, 62 homes.

Slater.
Josephine County. 34,000 acres (Oregon); 700 homes, 2 deaths (California).

South Obenchain.
Jackson County. 32,700 acres, 33 homes.

Thielsen.
Douglas County. 10,000 acres.

GIS Mapping by Teresa Zena Alcock, Oregon Department of Forestry, November 18, 2020.

Statistical data courtesy of Mason, Bruce & Girard, December 8, 2020.

Map 1. Western Oregon 2020 Labor Day Fire Perimeters, Names and Statistics.

Figure 2 shows rural cabin sites along the McKenzie Highway; among dozens that burned during the Holiday Farm Fire. Many of these were prized riverfront properties that had been within families for generations. All had a desirable market value. Now it appears that few of these structures may ever be rebuilt due to changed zoning laws and riparian regulations since they were first built and sold. This has effectively reduced the value of many of these properties to nearly nothing and made possible reconstructions highly problematic.

The large majority of homes that burned in the fires were aging trailer houses, mobile homes, and RVs, closely spaced together in commercial parks along state highways. Figure 3, for example, shows the results of the Almeda Drive Fire on the Dun Rov-N RV Park: one of many such parks that burned between Highway 99 and the Bear Creek Greenway.

Maps 2 and 3 show the locations of numerous other pre-1983 trailer parks that burned in the Almeda Drive Fire.

They also illustrate the effectiveness of Highway 99 and Interstate 5 in acting as fuel breaks, as ready access to fighting the fire, and in keeping it mostly confined to those boundaries. Ironically, a primary fuel along Bear Creek, other than homes, was invasive weeds -- principally Himalayan blackberries, which could have been readily managed with little cost but may have proliferated due to riparian management restrictions.

Fig. 2, Burned cabins near Finn Rock, north bank of the McKenzie River. Holiday Farm Fire aftermath, October 6, 2020 video-clip by McKenzie Peters, NW Maps Co.

Reforestation Planning

Reforestation planning is based on ownership. Implementing those plans is largely dependent on access. The goals of any plan are also restricted by existing and emerging laws and regulations, and by economics. Understandably, federal and state lands put laws and regulations as a first priority, while private landowners necessarily put economics first. The principal budgetary difference is whether taxes are being paid or being spent.

For private landowners, timber is the primary crop for making an income and being capable of paying taxes; on public lands it is usually more important to consider wildlife habitat regulations and public recreation and to involve the public in the planning process.

Reforestation planning, then, for private lands, can be little more than following the established industrial practices of large, successful timberland owners, such as Weyerhaeuser or Georgia-Pacific. The problem is when the resulting "tree farms" are burned before reaching maturity, such as occurred during the Labor Day Fires. How could that have been prevented?

Fig. 3. Burned trailer park on Highway 99, west bank of Bear Creek. Almeda Drive Fire aftermath, October 26, 2020 video-clip by McKenzie Peters, NW Maps Co.

On public lands the problem becomes a lot trickier. Even modest salvage plans are likely to be legally challenged by environmental organizations until the burned timber becomes useless or greatly devalued. Roads may be left closed or even removed on purpose. Herbicide use may not be allowed to control weeds, and trees may be kept from being planted for some reason or another. What to do?

Planning on public lands becomes even more complex when professional "stakeholder" groups and their legal representatives become involved. The great costs and marginal results of these processes are well documented. Many of the large-scale wildfires of the past 30 years have predictably resulted from these efforts. Something as simple as replacing a few culverts after several years of meetings and studies can be declared a "success." You could look it up.

The goals of reforestation planning, then, can be measured in jobs, profits, wildlife populations, and recreational visitor days -- and aesthetics, clean water, clean air, and public access. Wildfire mitigation must also be included.

In recent years the public focus on forest management has been on profitable timber production and wildlife habitat designations. The objectives of the former are obvious: grow as much merchantable timber as efficiently as possible. The latter concern has largely replaced 20th century concepts of "multiple use" on our public lands; and often specifically refers to habitat creations containing large, older trees, understory trees, snags, and dead wood on the ground. And few jobs.

Recent events have shown that these methods have not been working. The Labor Day Fires burned nearly as much timberland as designated habitats. Recreational uses were ruined, people and animals killed, and homes lost. Now would seem an excellent time to consider alternative approaches -- maybe even abandoned methods that have proven successful in the past.

When possible, reforestation planning can best be considered at a subbasin scale: that is, the land that is visible from ridgeline to ridgeline and is defined by a primary creek or stream. Fish spawn and birds nest within subbasins; insects, reptiles, amphibians, and small mammals rarely leave them; roads and trails within their boundaries provide systematic access, or not; and both wildfires and prescribed fires respond predictably to their individual fuel loads and firebreaks.

Map 2. Annotated 1983 USGS map with trailer parks along Bear Creek.

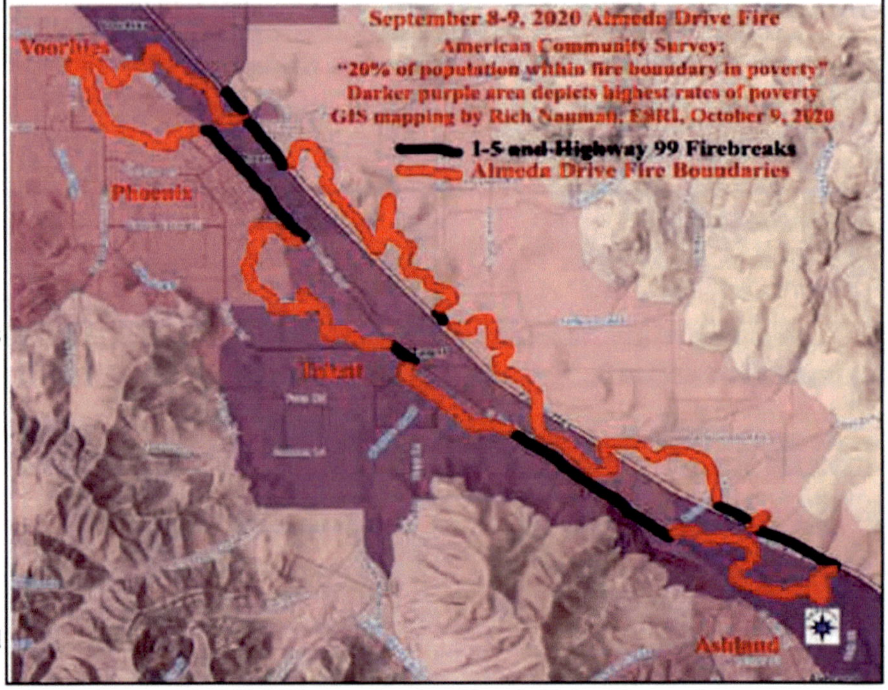

Map 3. Annotated 2020 GIS map showing Almeda Drive Fire perimeter.

In precontact and early historical time there were many open areas established and maintained by people in nearly every western Oregon subbasin. Typically, villages and campgrounds were located at the mouths of major streams and rivers and well-traveled trails followed the streams and their ridgelines. Tinder and firewood were regularly gathered and used along these routes and locations, and edible roots, bulbs, greens and berries were often established and harvested in adjacent fields and meadows. In essence, a systematic network of fuel-free campgrounds and trails formed major firebreaks along nearly every stream and ridgeline for hundreds or thousands of years prior to white immigration.

Could modern day replication of these conditions be useful for mitigating future wildfires and also for creating stable habitats for native species that had previously adapted to those plants and patterns over millennia? Impacts on timber production, recreation, and aesthetics would also be positive, and needed year around jobs and additional forest products would be created.

Another key difference was what foresters call "stems per acre." In past times it was common for Douglas fir trees to be spaced much wider -- and over far fewer acres -- than today. Industrial foresters commonly plant 300 to 700 trees an acre in hopes of getting 100 or so to a commercial size in areas where only 10 to 30 trees an acre formerly grew. With today's high quality growing stock and with better-quality site preparation, we should be able to plant far fewer trees, bulbs, and cuttings than in the past.

Site Preparation

Salvage logging, where possible, is the key first step in preparing a burned area for reforestation. This process is usually profitable and also performed in order to remove unsightly, flammable, and dangerous snags, which are often responsible for expensive reburns, road damage, and erosion.

Whether or not salvage has taken place, there are usually massive amounts of dead and drying fuels that remain: limbs, rotting logs, sub-merchantable trees, shrubs, duff, and flash fuels such as dry grasses and dead leaves.

These hazards can be eliminated through mechanical brush piling on flat and sloping lands and by prescribed cutting and broadcast burning on steeper lands. Brush piles can be safely burned in winter and spring while broadcast burns do best -- with the least damage to native plants -- in late summer and early fall.

A second problem -- and an opportunity -- created by wildfire is the exposure of unwanted weed species, such as exotic grasses, Himalayan blackberries, tansy, and scotch broom. These plants are generally located and spread along roadsides and streams and are rejuvenated by greater amounts of sunlight, the removal of competing vegetation, and the added nutrients following a fire. Such exposure makes them very vulnerable to spot spraying with herbicides for precise control before they seed or otherwise spread, rather than considering more costly and controversial aerial applications.

A well-prepared site is cheaper and easier to reforest and maintain and is more likely to be profitable and useful over the life of the next forest, however long. If a site is not properly prepared it is also more likely to burn again, and with greater damage, as clearly documented by the 1902-1928 Yacolt fires, the 1933-1951 Tillamook fires, and the 1987-2018 Kalmiopsis Wilderness fires. By contrast, there is little record of well-prepared sites being subjected to such reburns.

Tree Planting

Once a subbasin has been successfully weeded and mostly cleared of standing and ground fuels, no matter the time involved, it is ready for planting and seeding; either to safely replicate past habitat patterns or to focus on timber production and recreation. Wildfire risk has been greatly reduced and site productivity -- whatever the measure -- has been greatly increased by this process.

In western Oregon planting is accomplished almost entirely with Douglas fir seedlings. These trees typically grow in nearly pure even-aged stands from slightly inland of the Pacific coast to the higher elevations of the western Cascades. For this reason, forest wildlife conditions for the past several thousands of years have been largely defined by these trees, whether as even-aged seedlings, saplings, second-growth, or old-growth -- which also defines some of the most profitable forestlands in history.

In the past, stands of Douglas fir developed after wildfires, volcanic eruptions, windstorms, clearcuts, and plantations. In early historical times these stands were bounded by oak savannahs, huckleberry fields, open ridgelines and streamsides, camas meadows, southern balds and grassy prairies. Now these latter environments have been largely replaced with cities, farms, highways, riparian buffers, and expanded stands of more densely spaced even-aged Douglas fir.

With redesigned reforestation strategies, other species and habitats, such as open roads and trails, sunny meadows, cedar stands, ash swamps, and myrtlewood groves might be good to consider. Tree planting projects could be enhanced by the simultaneous planting of native root plants, such as camas and iris; shrubs, such as huckleberries and currants; and seeding with native bunchgrasses and wildflowers.

Assuming a site has been well prepared, and plans have been made regarding desired species and their locations, the next things to consider are planting specifications and stock qualities.

In the past, tree seedlings were often planted in grids at precise eight to 12-foot distances, resulting in anywhere from 300 to 700 trees per acre. Mortality, precommercial thinning, and occasional commercial thinnings typically reduced this number to less than 60 or 90 trees per acre at time of harvest.

This great amount of costly mortality was often caused by poor-grade site preparation and/or low-grade seedlings to begin with, and poor-quality planting and stand maintenance standards to follow.

Without going into technicalities, planting stock often consisted of one- and two-year old seedlings lifted straight from the nursery and often poorly grown or handled. Large numbers of these seedlings -- sometimes all of them -- would die after planting and it would be necessary to replant the area again. Seedlings that were grown for two years and then transplanted and root-pruned an additional growing season ("2-1s") did far better but were less common and more costly to grow and plant.

In the last several years of my reforestation career, we began favoring "plug-1" seedlings for uniformity and quality and the fact that, being initially grown in a greenhouse, they could be obtained in less than two years. With a constant 90% to 100% survival rate for the plug-1's and a shift to "micro-site" planting densities, we were able to significantly reduce per-acre planting and subsequent stand maintenance costs -- and the trees grew larger, faster.

"Micro-site planting density" involves selecting the best available location for an individual seedling within a given radius. For example, to plant 150 trees within an acre, a planter would place the seedlings in the best locations -- such as the downhill side of stumps or deep mineral soil above a stream's highwater mark -- every five to 20 feet at an average 16 or 17 feet apart. It is easier to learn than describe.

Active Management

After a burned forest area has had its roads opened and

power restored, any potential salvage completed, site prepared, and then planted, the results will still have to be actively managed in order to achieve long-term management goals. This is true whether the intent is to grow timber, create recreational opportunities, and/or develop wildlife habitat.

Roads and trails will need to be maintained in order to retain commercial and recreational access to affected lands, and also to help address any future wildfires, windstorms, insect outbreaks, noxious weed infestations, etc.

Fig. 4. Gordon Meadows with Bob Tom, Don Day, and Wayne Giesy, June 17, 2007. This area formed an ancient and effective firebreak in the South Santiam headwaters. It was regularly visited and occupied by Molallan Indians for many years in order to hunt, trade, and to harvest camas, nearby huckleberries, and beargrass. Deer, elk, bears, cougar, butterflies, and songbirds would have also been regular visitors, due to available food and water. Nearby forestlands had far fewer trees and would have been regularly scoured for firewood for camping, cooking, drying and camas baking purposes. For the past 500 years, most trees would have also been smaller and younger. The restoration of early historical Molallan villages, campgrounds, camas meadows, huckleberry fields, and trail networks would produce significant local work and income for many years, greatly reduce wildfire risk and severity, be far more attractive by most standards, and would be a much safer environment for people and wildlife. Photo by the author.

Dead and heavily damaged seedlings may need to be replanted and selective uses of herbicides and animal control measures might be necessary. Spot spraying of herbicides around individual trees, rather than aerial applications, is often a cheaper and more effective process. This uses a minimum of chemical, little fossil fuels, and leaves significant alternative browse available for animals that might otherwise favor the seedlings, such as deer, elk, mice, or rabbits.

After seedlings have become well established, precommercial thinning of stems to desired commercial or habitat densities become important options. Thinning when seedlings are only five or six years of age is a lot cheaper, safer, and more effective than typically waiting until they are 15 or 20 years old. The earlier thinning also reduces the time and amount of dead wood and foliage on the ground and keeps areas of shrubs, grasses, and wildflowers open between the trees.

Prescribed burning can be done to rejuvenate berry fields and grasslands, to dispose of brush piles, and/or to keep roads and trails open, as people have done for thousands of years. Traditional burning practices also greatly reduce the likelihood of damaging wildfires.

If these strategies are generally followed, the resulting forests will continue to provide safe, quality habitat for a wide range of species, greatly reduce risk and severity of wildfires, and timber crops will reach merchantable size and better quality much sooner.

The differing landowners affected by the Labor Day Fires will necessarily have different approaches to reforesting their properties, as described. It will be interesting to see what they do and what the results will be five and 10 years from now.

The Snags of the North Umpqua:
Part 2. Forest Restoration

By Bob Zybach, Ph.D

Figure 1. 2021 Jack Creek Fire plume, burning through snags remaining from the 2002 and later Apple Fire(s), stopping traffic along North Umpqua Highway 138, July 8, 2021. Video-clip by McKenzie Peters, NW Maps Co.

This is the second of two articles regarding the risks and mitigating strategies associated with the thousands of acres of snags remaining from major forest fires along the North Umpqua River during the past 20 years.

My previous article focused on the short- and long-term risks associated with leaving the snags in place. In particular, risks were considered regarding local residents, firefighters, travel and power corridors, private properties, wildlife populations and recreation.

The risk of snags in the environment is well documented. Depending on their size, age, and distribution, they pose a constant danger of burning in a wildfire. They can also cause death and serious injury when they fall, damage roads and buildings, host harmful insects and diseases and, to many people, are unsightly blemishes on the landscape.

Their greatest danger, though, is their flammability in a wildfire and the deadly and costly destruction that can directly result.

BACKGROUND

The September 8, 2020 Labor Day Fires in western Oregon burned nearly a million acres of land in a three-day period, killed 11 people, destroyed more than 4,000 homes, polluted the air with toxic smoke for nearly two weeks and killed millions of native wildlife.

Of these amounts, the Archie Creek Fire, along the North Umpqua River, was responsible for one human death, more than 150 people losing their homes, and over 131,000 acres, mostly forested, being burned. Due to the fire's large size and rapid spread, mortality of native plants and animals approached 100% within much of the fire's perimeter.

Following this event, the 2020 USDA Rapid Assessment Team (RAT) noted:

"Over the past 20 years, 28% of the Umpqua [National Forest] has burned in wildfires, with the total acreage being higher due to several areas being burned two to three times over the past 20 years. Less than 1% of these past fires in total have been salvaged . . ."

An important statement in this quote is that "several areas . . . burned two or three times over the past 20 years." Following the RAT report, the 2021 Jack Fire, Chaos Fire and Rough Patch Complex have burned more than 70,000 additional acres and the total area burned in the North Umpqua basin since 2002 is now more than 300,000 acres. Two-thirds of this amount occurred in just the past two years.

Map 1 and Table 1 show the great number and large size of major wildfires that have occurred along the North Umpqua during the past 20 years. Almost all of these fires have taken place entirely within the boundaries of the Umpqua National Forest (NF), where fewer than 1% of the dead trees have been salvaged. (Note: Both maps used for this article were developed by the author and by Vyla P. Grindberg, Douglas County GIS Coordinator.)

Figures 2 and 3 show the increasing hazard to rural urban areas posed by these fires. In addition to the increased flammability of public forestlands caused by the accumulation of snags, growing green fuels have also been directly connecting these lands to nearby homes and businesses. Note the increasing volumes of trees and shrubs along the Umpqua River, within town, and moving in from hillsides that had been burned or grazed in earlier times.

Also note the ever-increasing number of closely spaced, highly flammable, wooden structures that have been built in recent years. The Labor Day Fires destroyed more than 4,000 homes, and the majority of these losses were also fueled by adjacent hillside and riparian vegetation that had developed uncontrolled during the past few decades.

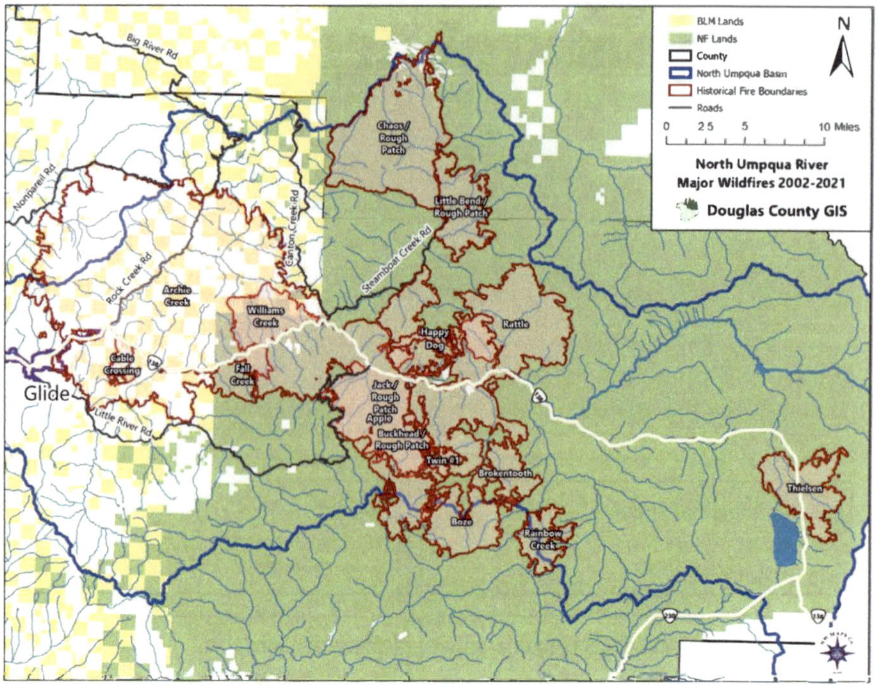

Map 1. Major Forest Fires of the North Umpqua River Basin, 2002-2021. Note northern boundary of 2021 Chaos Fire as closely following ridgeline separating Willamette and Umpqua River basins. Similar patterns can be shown for much of the perimeters of the 2009 Williams Creek and 2020 Archie Creek Fires.

REFORESTATION PLANNING

Following a major event in which widespread deforestation occurs, whether from wildfire, windstorm, clearcut, volcanic eruption or other occurrence, an important next step is to make specific plans for both short- and long-term futures. In an ideal world, those plans would directly include local businesses, residents, and students.

Several reasons exist to best consider restoring devastated forestlands on a subbasin-scale basis. For purposes of this article, subbasins are defined as the individual areas drained by named creeks tributary to the North Umpqua River. Most, if not all, of these streams are fish-bearing. Subbasin considerations include historical fire behavior patterns, endemic plant and animal populations, enhanced human safety, existing road and trail patterns, and land ownership.

Table 1. Historical North Umpqua River Wildfires, 2002-2021.

Year	*Date*	*Wildfire Name*	*Acres*	*Snags*	*Ownerships*
2002	0816	Apple	17,600	Yes	USDA Umpqua NF
2008	0813	Rattle	19,800	Yes	USDA Boulder Creek Wilderness
2009	0912	Boze	10,600	Yes	USDA NF
2009	0912	Rainbow	6,100	Yes	USDA NF
2009	0728	Williams Creek	8,400	Yes/No	USDA NF/Private
2015	0728	Cable Crossing	1,900	Yes/No	USDI BLM O&C/Private
2017	0808	Fall Creek	4,800	Yes	USDA NF/USDI BLM
2017	0808	Happy Dog	31,400	Yes	USDA NF
2017	0808	Twin #1	1,400	Yes	USDA NF
2017	0808	Brokentooth	3,700	Yes	USDA NF
2020	0908	Thielsen	9,800	Yes	USDA NF
2020	0908	Archie Creek	131,500	Yes/No	USDA NF/USDI BLM/Private
2021	0705	Jack Creek	24,000	Yes/No	USDA NF
2021	0730	Chaos	28,800	Yes	USDA NF
2021	0730	Buckhead	7,200	Yes	USDA NF
2021	0730	Little Bend Creek	9,400	Yes	USDA NF

The ridgelines separating named subbasins typically form natural firebreaks, as shown by Figure 4. Ridgelines are also often the locations of ancient foot trails, logging roads, and/or highways, which can directly aid in modern access needed to actively manage lands and fires.

Too, subbasins are lifetime homes to most native plants and animals within their bounds; including trees, shrubs, wildflowers, reptiles, amphibians, small mammals, fish, and even birds, if hatchlings and nestlings are considered. Whether they are being managed as homesites, for timber production, recreation, and/or as wildlife habitat, subbasins have proven effective at both planning and operational project scales.

Legal planning constraints, rather than physical or biological, are based almost entirely on property ownership -- in this regard, the Labor Day Fires affected four basic ownership types: private timberland owners; USDA Forest Service; USDI Bureau of Land Management; and residential, including transportation and powerline corridors.

Map 2 illustrates how the Archie Creek Fire affected each of these types of ownership in relation to the named subbasins that were burned. For the most part, subbasins were entirely contained within Umpqua NF land to the east and private forestland to the west. Between these extremes, the burned Rock Creek subbasins are a mix of private and BLM O&C Lands.

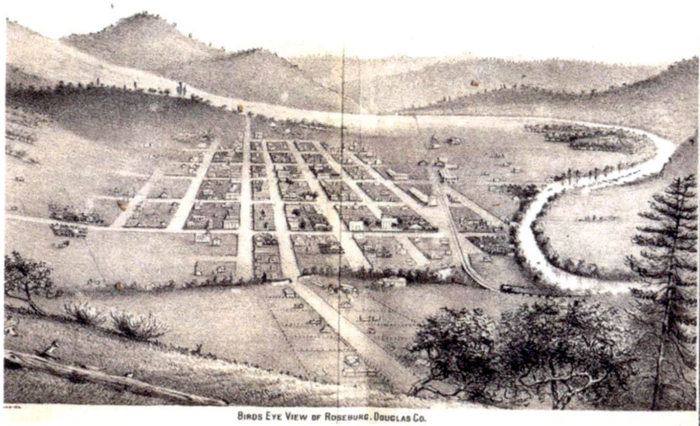

Figures 2 and 3. Roseburg, Oregon in 1885 and in 2009. Both perspectives are "from a rocky ledge" looking south: the drawing is from Albert Fagan's 1884 "History of Southern Oregon," and the photo is by BLM Forester, Abe Wheeler.

The large green polygon on Map 2 includes all three ownership types, as well as commercial and residential properties, State Highway 138, and the primary power lines corridor. For these reasons it is highly recommended that all snags be removed or carefully managed in these subbasins. This is where they are the most dangerous to people and property and where future landscapes featuring shrubs, native grasses, and/or wildflowers might be safer and more attractive.

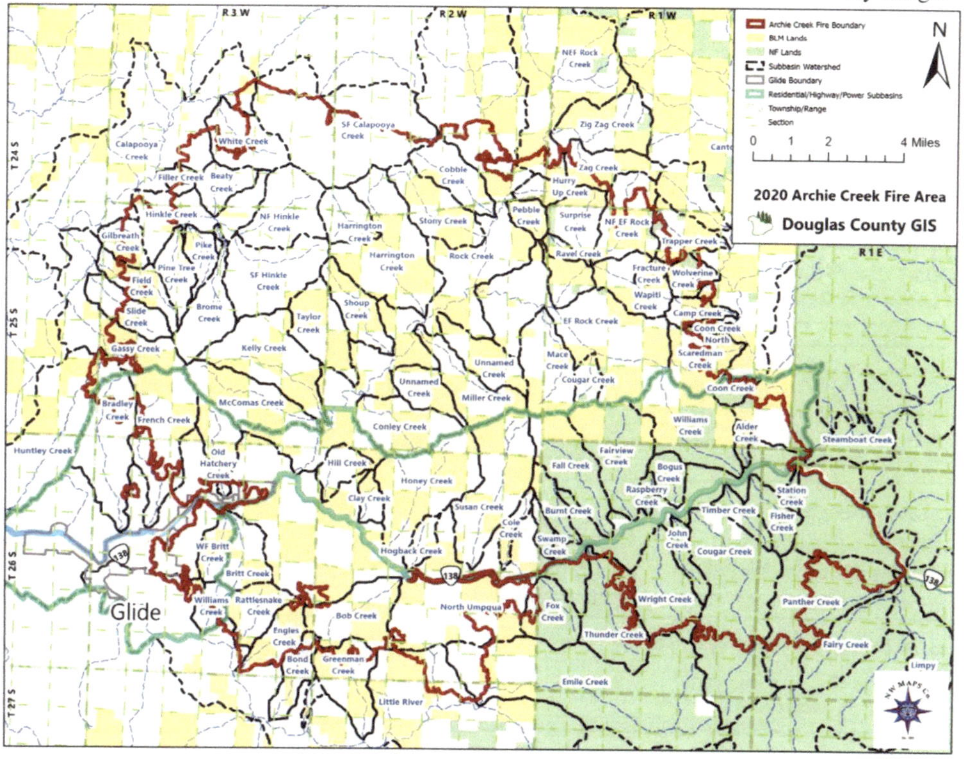

Map 2. Named sub-basins and land ownership patterns of the 2020 Archie Creek Fire. Note large green polygon including residential properties and major transportation and power line corridors.

For private forest landowners, timber is usually the principal crop for making an income and being capable of paying taxes; in recent years on public lands, it is usually more important to consider wildlife habitat regulations and public recreation and to involve the public at some level in the planning process.

The common goals of reforestation planning, then, can be measured in jobs, profits, wildlife populations, and recreational visitor days -- and safety, aesthetics, clean water, clean air, and public access. Wildfire mitigation must also be included.

Implementing those plans is largely dependent on physical access and funding. The goals of any plan are also restricted by existing and emerging laws and regulations and by economics. Understandably, federal and state lands put laws and regulations as a first priority, while private landowners necessarily put economics first. The principal budgetary difference between ownerships is whether taxes are being paid or being spent.

SITE PREP

Prior to beginning planting, seeding, or other forest regeneration practices, it is important to do something regarding residual fuels and unwanted competitive vegetation. This process is often referred to as "site preparation" or "site prep."

Immediately following a fire, roads are often blocked by fallen (and falling) limbs and trees and travel can be very dangerous within the area until these problems have been resolved. A new tool for dealing with these situations is the video drone (Figure 5). An operator can guide the drone along affected trails, roadways and streams at eye-level and produce a very useful permanent record for road maintenance, potential timber buyers, forest managers, scientists and interested public.

The use of drone videos is safer, much cheaper and can provide far more detail than visiting the area by vehicle or observing it from the air. Further, drones can be programmed to travel identical routes over time, providing an excellent method for periodically monitoring projects, changing road and trail conditions, floods, landslides and/or forest growth and mortality.

A large-scale wildfire usually leaves more fuel in its wake than it consumes. In addition to snags, burned brush and logs, these fuels often include undesired weeds such as Scotch broom, Himalayan blackberries, tansy ragwort and exotic grasses that can be rejuvenated by fire. For this reason, a new burn is often the best condition for locating and treating weeds as they emerge, and before they go to seed.

Figure 4. 2020 Beachie Creek burn pattern, showing abrupt burn boundary along a "natural firebreak" ridgeline boundary to the fire, January 18, 2021. Drone photo by Andy Adkins, Floating Point Films.

One of the most efficient methods of preparing a post-fire subbasin for planned reforestation is called "slash and burn." This process uses prescribed fire to remove flash fuels, treated weeds, ground fuels, ladder fuels, and other flammable debris

Figure 5. Opening sequence to 14-minute instructional video prepared for spring-term Southwestern Oregon Community College forestry class. Published June 27, 2019 by the author and by Portland videographer Frank Mahoney.

such as snags, tops, limbs and twigs from an area prior to taking other actions. This method was developed decades ago, as illustrated by Figure 6, and has proven to be a highly effective process for ensuring worker safety and for planting quality and success.

"Slashing" refers to the practice of uniformly cutting snags, shrubs, limbs and other unmerchantable plant materials to less than a two-foot-or-so level over an entire area scheduled for regeneration. The area is then "broadcast burned" to remove as much of this material as possible. If fuels include significant numbers of green trees and shrubs, they can be allowed to dry, or desiccated with chemicals, before burning. This process can further improve project efficiency and quality and also helps reduce visible smoke and smoldering from the fire.

Unwanted residual plants and materials remaining from a wildfire can also be "mechanically treated" on flat or sloping ground with a Caterpillar or other equipment. Typically, a toothed blade is used to uproot, pile and/or "windrow" surface and subsurface fuels, leaving mostly bare soil behind for planting or seeding.

While piles and windrows of concentrated fuels are usually burned during winter or spring when wildfire risks are minimal; broadcast burning is often best done during the summer and fall fire seasons. Not only are fires more effective at clearing unwanted fuels during these times, but hotter burns also result in better smoke management conditions. Another value is that thousands of generations of native plants and animals have adapted to this timing and often even benefit by this practice.

TREE PLANTING

Short-term planning must reasonably focus on clearing roads and trails, conducting any commercial salvage operations, and preparing the area for planting or other long-term uses. Most private and state forestlands require planting for deforested areas, meaning that reforestation planning must also include considerations of quality, quantity and species of seed and seedlings to be used.

One of the major reasons for the unusual recent burning of managed tree plantations is the contiguous canopy of pitchy, flammable conifers that are formed in accordance with State regulations and timber markets. When an area is first burned

Planting a broadcast burn on the Oregon Coast - 1979

Figure 6. Lincoln County tree planting unit that had been commercially logged, slashed, and burned prior to planting. Note the number and size of unmerchantable logs that had skidded to the bottom of the draw after being felled, and the relative size of the workers. This successful plantation, near Elk City, is now likely coated with 45 year old timber or has been logged and planted again. Photo by author.

Figure 7. April 1952. Handwritten notes on back: "Group of 300 High School kids from Elkton and Reedsport doing their annual tree planting program -- on private and BLM timberlands -- this year on lands burned by the large Vincent Creek-Weatherly Creek Fire of 1951. Gathering for their lunch of hot dogs, ice cream, and sandwiches. (Led by Foresters from Coos Fire Patrol, Douglas Fire Patrol, and BLM.)" Photo and notes courtesy of Jerry Phillips, Coos Bay.

or logged it might contain a dozen or maybe 50 or 60 merchantable conifers per acre; often growing in clusters and strips with open ridgelines. It is then quickly planted with hundreds of trees per acre, usually in a solid grid-like pattern of Douglas fir seedlings, to meet economic goals and State regulations.

Figure 7 shows a large planting operation in western Douglas County involving local school children in the early 1950's. This was near the beginning of hand planting logged or burned areas, and most work at that time was performed by local contractors and residents. As a student in the early 1960's, my Portland high school biology club spent a Saturday planting trees on the Tillamook burn. These plantations were intended to create future jobs and a lot of pride and care went into their establishment and growth.

The Archie Creek Fire subbasins shown on Map 2 form some interesting general patterns: the Calapooya Creek subbasins are mostly private, industrial forestlands; the Rock Creek subbasins are mostly composed of the O&C "checkerboard" forestlands; and the mainstem North Umpqua subbasins are a mix of residential, industrial, O&C Lands and Umpqua National Forest. Each type requires road access to repair and restore damaged properties, and each type has differing laws and regulations regarding what can, or must, be done.

Most private properties affected by the 2020 Labor Day Fires fire are industrial timberlands that have already been salvage logged and/or even planted, depending on their condition when they burned.

Historically, industrial tree plantations have been mostly safe from wildfire, but Rock Creek tree farms are checkerboarded with BLM Lands which may continue to contain large amounts of snags. At some point these snags are likely to fall or spread wildfire on to adjacent private lands or access roads, which will remain a principal risk to these properties.

An alternative approach might be to mimic early historical forest patterns. These could feature open ridgelines and riparian areas, with native hardwoods, such as oak, ash, alder, myrtle and cottonwood, in key locations. Restored camas and beargrass meadows, grassy prairies, and huckleberry fields have proven recreational and commercial potentials and can also serve as effective firebreaks during wildfires or controlled burns.

Instead of planting trees in eight-, 10- or 12-foot grids (300 to 660 seedlings per acre) and then precommercially thinning them to 150 to 200 trees per acre after 10 or 15 years, it makes greater economic sense to only plant 150 or 200 trees in the first place. This can be readily accomplished by using "micro-site" densities to determine planting locations and using high-quality seedlings that can be grown in less than two years and typically have survival rates exceeding 90 or 95 percent.

These results could be achieved without significant reduction in area devoted to timber growth -- and which can produce desired product volumes with far fewer plants per acre.

SITE MAINTENANCE

A functional reforestation plan should include a long-term maintenance strategy involving regular monitoring, disease and pest control, occasional pruning, thinning, weeding and situational salvage. Prescribed burning must also be considered for these purposes and clearly outlined in the plan.

Road and trail access is the most critical consideration in the active management of forested properties and is also important to consider regarding the management of future wildfires, floods, landslides, ice storms, snowstorms, windstorms, and their aftermaths. Of most strategic importance are roads and trails following ridgelines and streambanks.

Functional roads and trails are among the most valuable assets we own and enjoy on our public lands. These routes have historically followed streams and ridgelines, which further serves to reinforce their value as firebreaks, as well as providing common residential, work, and recreational access.

Good rock roads may be gated on industrial timberlands to reduce vandalism, while on federal lands they may be entirely abandoned or even made inaccessible on purpose. The same situation exists for recreational trail networks on public forestlands. Managing the predictable large-scale wildfires on federal properties is often made more dangerous, difficult, costly, or even impossible, for this reason.

Forests of dead trees are far more flammable, dangerous, and unsightly (to most) than forests with living trees. Dead trees quickly become air-dried firewood, and dead forests become likely firebombs that almost certainly will burn again and again, unless something is done to stop that process. The easiest method of reducing or eliminating this risk, and occasionally the most profitable, is to remove snags periodically as they occur.

MORE RESEARCH

The 2020 western Oregon Labor Day Fires, including Archie Creek Fire, were historically significant for a number of reasons. These included the record number of acres burned, the consistency of affected major land ownership types, and the great diversity of fuels, elevations, slopes, and aspects to consider.

Another common denominator was the unusually heavy east winds, low humidity, and high temperatures that blanketed western Oregon and drove these fires. This combination of simultaneous events, common weather patterns, and mixed ownerships and management histories provides a great opportunity to study and compare current and future wildfire mitigation strategies.

Further research on these topics would certainly seem timely and useful, and the opportunity is both unprecedented and potentially very valuable to future forest owners and managers. Given the diversity of ownerships and the potential value of such findings to both public and private interests, perhaps a long-term collaborative research project along these lines should be seriously considered.

NOTE: This article is derived from a report contracted by Douglas Timber Operators with Dr. Zybach and NW Maps Co.

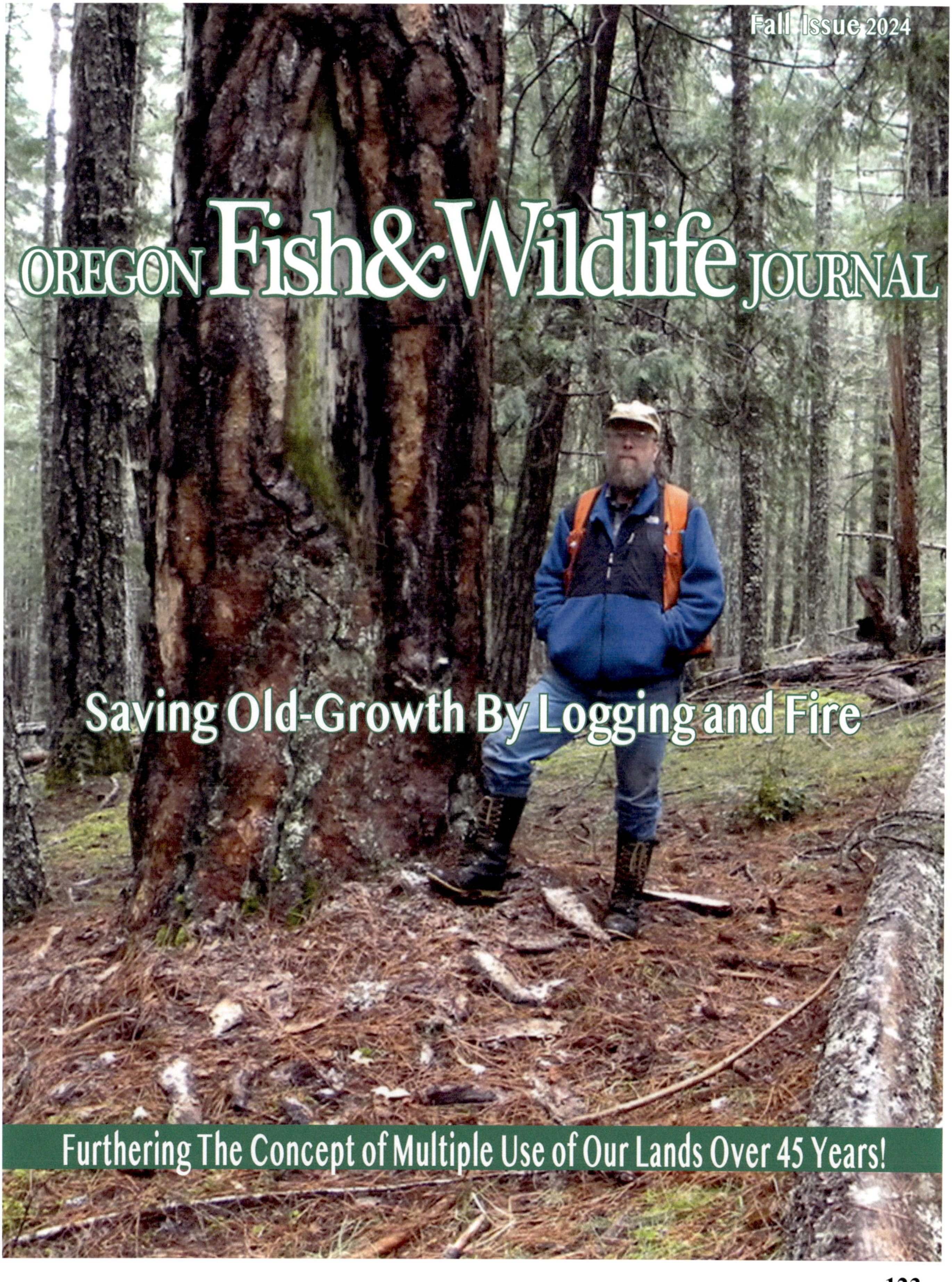
Fall Issue 2024
OREGON Fish&Wildlife JOURNAL
Saving Old-Growth By Logging and Fire
Furthering The Concept of Multiple Use of Our Lands Over 45 Years!

Jim's Creek Revisited: A Burning Opportunity

By Bob Zybach, PhD.

Fig. 1. Tim Bailey next to a "culturally modified tree" likely peeled last by a Klamath or Molallan Indian before 1855. It is important to remember that this tree was much smaller and younger at that time. There were about 100 modified pine in the Jim's Creek project, but about 1/2 had been killed by invasive Douglas fir trees, such as these, competing for sunlight, water, and soil. A crown fire or ground fire under these circumstances would be fatal to this tree. Photo by author, April 9, 2010.

Oakridge is walking distance south to a former 25,000-acre oak and pine savanna. The community is largely defined by western Cascade Mountain ridgelines at a little more than 1200-feet elevation. This puts it above the Willamette Valley fogs to the west, and lower than the heavy snows at higher elevations to the east, resulting in more than 300 days of sun a year.

The reason oak and pine were able to persist over such a large area was entirely due to hundreds or thousands of years of regular burning by local Indian families, who may have also established the original oak and camas populations throughout the area, and possibly even the pine and hazel. Otherwise, the hills and ridges would be almost entirely covered with even-aged Douglas fir trees, as with most of forested western Oregon.

Oakridge was the site of a Molala Indian village with a large camas swale that barely existed into the 1850s, when most of the remaining people were taken to the Klamath, Warm Springs, or Grand Ronde reservations. A few stayed behind.

Charlie Tufti, a well-liked and renowned Molala hunter, and historic founder of Waldo Lake, had a 160-acre Homestead on a hill to the east of the old village. His first wife was Klamath, and she and her baby died, as did his second wife, a Molallan, and their boy and young girl. At least seven people are thought to be buried on the land, and typhoid was the most likely cause of most or all of their deaths. His third wife was a Warm Springs woman named Lucy. In 1889, Tufti sold his ranch and they moved to Warm Springs and became prominent members of the Reservation, where they successfully raised their two children, Josephine and Jasper. Their descendants still have strong ties to the Oakridge Homestead to this time, more than 135 years later.

The Oakridge/Westfir village(s) was/were along a major east-west route from the Willamette Valley to eastern Oregon used by people for thousands of years; and along the historic north-south Klamath Trail trade network that connected the Klamath Lakes to Celilo Falls on the Columbia, to Willamette Falls, and to the Willamette and Umpqua valleys.

The main village site was not only an ideal seasonal trade location; it was also a short distance to more than 25,000 acres of what foresters call a "mixed conifer forest." That is, 25,000 acres of grassy white oak, yellow pine, sugar pine, Douglas fir, and incense cedar savanna, teeming with deer, elk, bear, and fish, with fields of the world's finest huckleberries and beautiful camas meadows. All created and maintained for thousands of years by people who were lifelong experts with fire, beginning from early childhood.

After the remnant Molallan and Klamath families died

Fig. 2. One of the few remaining large oaks in Jim's Creek. Most of the others, along with many old-growth Ponderosa and sugar pine, have been shaded out by invasive Douglas fir since the late 1800s. Photo by author, August 19, 2011.

or were moved to reservations in the 1850s and became replaced with white settlers, crops, and livestock, the burning stopped, and Douglas fir rapidly began filling the savanna grasslands with waves of even-aged seedlings.

The variously-aged early historical savanna trees, which often bordered open fields, existed as solitary landmarks in meadows, or formed scattered groves in grassy prairies, rarely numbered more than 10 or 20 an acre, and often less. In the absence of regular burning, even-aged seedings of Douglas fir soon began populating the area at an eventual rate of 150 mature trees per acre, according to foresters.

Douglas fir competes with other vegetation by growing taller and faster and creating a solid canopy that shades out most other plants, including younger fir trees, other tree species, shrubs, grasses, and wildflowers. Mushrooms and mosses benefit. By the year 2000, Douglas fir had killed most all of the oak and about one-half of the savanna's old-growth pine and was reaching a commercial size of 15 to 20 inches diameter. It had also shaded out most of the grasses favored by deer and elk, and the flowers favored by butterflies, hummingbirds, and people.

Following World War II, the adjacent communities of Westfir and Oakridge had both developed significant lumber manufacturing facilities, and the Oakridge Ranger District of the Willamette National Forest soon focused on the lucrative sale of hundreds of millions of board feet of public old-growth and mature second-growth Douglas fir that surrounded the former savanna. The home building and heavy construction boom that followed the war provided great jobs for the communities' loggers, sawmill workers, Forest Service employees, teachers, and businesses.

In 1953, Oakridge and Westfir held the first of 71 -- and counting -- annual "Tree Planting Festivals" to celebrate the successful reforestations of the ever-growing private and public clearcuts. New roads were being built into the forest, campgrounds constructed, and hunters, hikers, fishermen, and bird watchers benefited. Reforestation was being successfully achieved with industrial grid-like plantations of Douglas fir seedlings. The intent was to leave another lucrative crop of trees for future generations by growing as much valuable Douglas fir fiber per acre as quickly as possible.

Most of the lumber manufacturing operations moved away when federal timber sales began sharply declining -- largely due to environmental lawsuits -- and private lands had been mostly clearcut in the 1980s; in 1996 anti-logging protesters burned down the Oakridge Ranger District office; and in 2002 career USFS forester, Tim Bailey, and fire ecologist, Jane Kertis, formally proposed the "Jim's Creek Savanna Restoration Project."

Bailey had recognized that returning to the original open forest conditions would be a prime example of ecosystem restoration, which had recently become a Forest Service priority, and was also cognizant

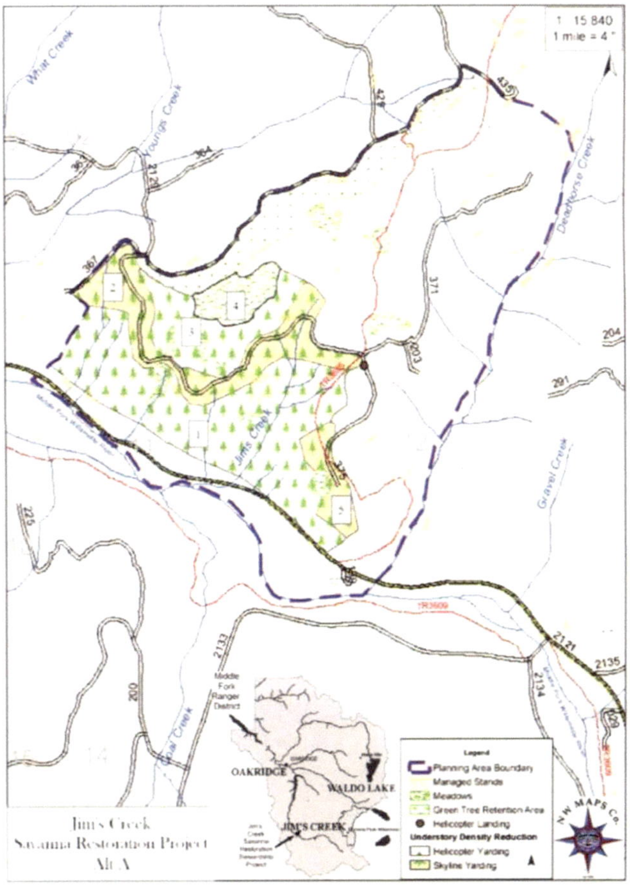

Map of Jim's Creek Savanna Restoration Project, 2005 USDA Forest Service Environmental Assessment, Alternative A.

of the fact that the tens of thousands of Douglas fir that had invaded the savanna were now approaching commercial size and could be marketed at a profit. More importantly, both were concerned about the ever-increasing loss of cultural landscape, and the directly related losses of oak, pine, native grasses, and the great diversity of songbirds, butterflies, reptiles, amphibians, small mammals, and large game animals that depended on these environments to survive.

Further, they were also concerned about the flammability of the contiguous Douglas fir canopy. A deadly crown fire had been all but impossible when the land was a savanna, but entirely typical of "stand replacement" wildfires in the region's even-aged conifer forestlands. This concern was borne out in 2009 when the Tumblebug Fire burned through 1,200 acres of the now-defined "mixed conifer forest" and killed many of the few remaining savanna oak, cedar, and pine in the area that had not already been shaded out. These trees had survived and prospered with dozens of purposeful fires intentionally set by people before 1855, only to die by wildfire 150 years later.

Fig. 5. Snags and dead, woody debris were not typical components of a savanna because of the regular burning and firewood gathering by people. Felling of dead pine trees and logging of invasive fir provided excellent opportunities to document the former savanna's previous fire history and abrupt ending of human burning and occupation. Photo by author, April 9, 2010.

Fig. 4. Tim Bailey and a portion of Jim's Creek that had been helicopter logged and then broadcast burned in 2009. Note the herd of elk near the crest of the reclaimed savanna and the deer to their lower left. A zoom lens counts 17 elk in the herd and three more and four deer along the timberline. Photos by author, April 9, 2010.

The proposed Jim's Creek restoration project was a little over 630 acres in size and fully intended to be a practical demonstration project for a much greater restoration effort with other public lands managed by the Forest Service in the ancient savanna. The property is 20 miles south of Oakridge on the east bank of the Middle Fork Willamette River and surrounded by roads and perennial streams, for good access and firebreaks, but contains a significant number of archaeological sites and scattered cultural resources that had to be considered.

Two major archaeological sites, Horse Pasture Cave and Vine Rockshelter, are just across the river from Jim's Creek. These locations date back thousands of years and contained hundreds of

Fig. 3. John Thomas and Tim Bailey in a grove of surviving old-growth on Jim's Creek. Photo by author, August 19, 2011.

Regulations vs. Restoration. Aerial photo of Jim's Creek project, July 13, 2012, following completion of logging operations. Note the strips of green that stretch into the open ground. These are 100-foot-wide "buffers" following ephemeral streams that are dry in the summer, have no fish, and never existed before in the history of this land. USFS regulations called for 300-foot buffers but allowed the lesser size following negotiation. These are atypical vegetation patterns in a savanna that increased logging costs, decreased income, and should have never been created in an area maintained for centuries by broadcast burning.

previously undisturbed artifacts when they were excavated in the early 1980s. Researchers discovered more than 400 obsidian arrowheads, basketry fragments, and numerous stone tools, including knives, drills, hammerstones, pestles, and anvils, along with significant amounts of bone from deer, elk, birds, and small mammals that were hunted. Horse Pasture Cave is near an open grassland, named Grassy Glade, that is still used for deer and elk hunting to this time.

These locations are just a short walk to the Jim's Creek campsite, on a flat near its mouth, and about five miles by ridgeline trails to the ancient huckleberry fields on Groundhog Meadows and Logger Butte, which are still being harvested by local families. The project area itself supports many species of edible plants, including camas, yampa, biscuit roots, oak, and hazel. Of particular note is the historic use of the inner bark of ponderosa pine trees, evidenced by the large rectangular scars where bark had been peeled more than 150 years ago. It is believed that the sap and cambium layer were used as food and other purposes. The practice was well known to the Klamaths, who regularly traded -- and often lived -- with the Molala. There is even speculation that the ponderosa and sugar pine groves had originally been created via seasonal Klamath visitors, and perhaps the oak and hazel by trade with Kalapuyans from the Willamette Valley.

About 100 such "medicine trees" are in the Jim's Creek

Fig. 6. Tim Bailey standing in a portion of Jim's Creek that should have been broadcast burned several years ago. Seedlings, shrubs, and large woody debris need to be reduced or eliminated to allow native bunch grasses, forbs, and shrubs to again define the meadows and understory. Photo by author, April 5, 2024.

project area, and about half have been shaded out by the younger, invasive Douglas fir. Pine rots fairly quickly, and most of these dead trees have already fallen over. Peeling scars are typically two to three feet wide and four to seven feet long. All have been tagged and inventoried and indicate Molala and Klamath people must have been present in the early spring, when the trees could be peeled, as well as late summer and fall hunting, burning, and gathering of huckleberries, acorns, filberts, and pinenuts.

Even under Bailey's expert leadership, the project still took a dozen years to complete. Following its conception in 2001, Bailey and Kertis' 2002 formal presentation and subsequent reports, an Environmental Assessment in 2006, a comprehensive "Watershed Analysis" in 2008, and signifi-

Fig. 7. Mature second-growth thinned in 2010 to begin savanna restoration efforts. Trees are top-heavy due to growing in close proximity and shading out lower limbs. Predictable windthrow has further thinned the stand and downed trees should ideally be removed before broadcast burning. Previously burned stump, native bunchgrasses, and deerbrush are in the foreground. Photo by author, April 5, 2024.

cant outreach and input by tribes, environmental groups, and the public, a logging contract was awarded to Swanson Superior to log the invasive Douglas fir in about 425 acres of the project area.

Because of the presence of the 100 "culturally modified" pine, eleven registered archaeological sites, and lithic scatters throughout the sale, a decision was made to log 70% of the area by helicopter to create as little surface disturbance as possible. This meant that harvesting costs were more expensive and, coupled with the poor markets at the time and helicopters only being available between wildfires, the contract took nearly four years to complete.

Fig. 8. An effort to eventually transform a pine plantation to its former savanna condition began with volunteer workers girdling a number of the young trees. As evidenced by this photo, some of the girdled trees are still living, and most of the others have fallen over and created a fire hazard. A broadcast burn "as is" would likely remove a lot more excess trees, but removing the fuel before burning would allow the trees to grow to a commercial thinning size to help pay future maintenance costs. Photo by author, April 5, 2024.

The logging resulted in 9.6 million board feet of commercial timber harvest and generated, "before adjustments," $660,000 in stumpage receipts; some $300,000 of which was earmarked for subsequent restoration activities by using the Stewardship Contracting authority granted by the Forest Service Regional Office after a year of review. The Middle Fork Willamette Watershed Council was also involved in the Jim's Creek project from "early on," and was "instrumental in helping to determine the priority of retained receipt expenditures." Naturally, they were also well represented in budgeting requests.

By July 2011, most logging and a few prescribed burns had been completed, so a successful public field trip was held that was attended by more than 30 people, including environmental group representatives. The tour generated widespread concurrence regarding the positive nature of the project's progress and results, and a nearly unanimous opinion that the savanna restoration concept should be greatly expanded, and with some urgency.

The management plan called for prescribed burns every five to ten years, but has never been completed. Instead, a year later, in July 2012, the Rigdon Dry Forest Restoration Project was being developed to expand the successful Jim's Creek pilot demonstration to a "considerable portion of the 28,200-acre Middle Fork Mixed Conifer Forest type." Further, "this forest type has changed dramatically over the last 100 years, due primarily to fire suppression and, (in the case of the 10,000 acres of plantations in the forest type) secondarily due to harvest activity."

There were some problems with this idea: in addition to the 10,000 acres of Douglas fir plantations, the project included 7,000 acres of private lands, more than 7,000 acres of "spotted owl habitat," another estimated 7,000 acres "in remote areas that would likely include the "taking" of spotted owls," and "the remaining 4,000 acres are populated with regulated tree voles."

••••••••••

I first wrote about the Jim's Creek project in 2012, as its completion was being delayed by environmental lawsuits related to the existence of hundreds or thousands of red tree voles that had multiplied with the Douglas fir encroachment. The self-described "environmental activists" were trying to stop logging by using this particular rodent, which had been recently determined by the Northwest Forest Plan (NWFP) to be a "species of interest." This federal recognition qualified the animal for "survey and manage" regulations, which had then become required by law. The voles weren't even listed as "threatened or endangered," and I wrote:

"This type of work stoppage, based on relatively new federal regulations and related litigation initiated by environmental organizations, has become the main difficulty in beginning and completing forest restoration projects in Oregon and throughout the West."

A local "Rigdon Collaborative Committee" was formed in 2016 to "work with the Forest Service on the Rigdon Landscape Restoration Project." The group of largely volunteers did a good job or organizing local people, business groups, and special interests through field trips and meetings to devise a formal strategy for overcoming these problems. A focus was on expanding the Jim's Creek model and doing it with local businesses and workers.

An initial 2017 project "restored" a 40-acre wetland, in part by placing logs in a creek, but by 2019 the collaboration had developed the Youngs Rock Rigdon project proposal. This ambitious effort identified 6,600 acres of desired forest management treatments to continue and expand the Jim's Creek restoration model. The strategy included a mix of "thinning, early seral creation, fuel reduction, habitat restoration, system road management, and recreation management."

A "Notice of Intent" (NOI) was published in the Federal Register in June, 2019; A "Draft Environmental Impact Statement Notice of Availability" (DEIS NOA, really) for allowing public comment was published in July, 2021; the "Final Environmental Impact Statement" (FEIS) NOA was published April, 2023; the "Objection Period" was started on the same day, April 7, and on May 18, Oregon Wild filed an objection, because of logging and global warming.

Fig. 9. Ground fuels and three different species of conifer seedlings that need to be treated. Photo by author, April 5, 2024

Further, they used the Jim's Creek Project as their rationale, because it had "resulted in a significant blowdown event killing several hundred old-growth pine trees that the project intended to save, as well as significantly spread weeds where native wildflowers were intended to thrive."

The "Youngs Rock Rigdon Final Record of Decision" (ROD) was signed by the Willamette National Forest Supervisor on December 8, 2023, and work was estimated to begin by May 1, 2024. On June 17, 2024, a coalition of Oregon Wild, from Portland and Eugene, and WildEarth Guardians, from New Mexico, filed a complaint challenging the decision. It was their proven cut-and-paste strategy of claiming the Forest Service had "violated NEPA by failing to consider reasonable alternatives" and by not taking a "hard look" at the project's "environmental impacts."

As usual, the complaint argued the same semi-science concerns, claiming the Forest Service hadn't "adequately analyzed the project's effects on carbon storage, climate change, mature and old-growth forests, fire risks, road construction, and northern spotted owl habitat." They also demanded that the Forest Service immediately stop all "road construction and commercial logging until the alleged NEPA violations are addressed." Only lawyers, the litigants' well-paid staff, and agency bureaucrats will now be allowed to continue working on this project, if documented history is a guide.

Enough should have been enough 30 years ago. These people are destroying our rural communities and our nation's forests and wildlife, and the federal government appears to be overly compliant. While taxpayers keep footing the bill and breathing the smoke on both sides.

Jim's Creek needs to be completed. All 638 acres should have been burned 10 years ago, as planned and paid for. If people actually care about "critical habitat" and "biological diversity" -- which are legal terms, not science -- they should burn Jim's Creek now, this fall, when the Molala would have burned it. If the Forest Service can't do it, how about a local business, as in pre-spotted owl litigation times? Or the volunteer fire department? Prescribed fires are a lot cheaper and safer than wildfires, and this would be a great opportunity to publicly display that difference.

Jim's creek has good road and stream boundaries, fuel preparation could be done in a few days time, and trained crews are available as wildfires are being extinguished. Late summer and fall burns, as people have done for thousands of years, would greatly reduce risk and severity of wildfires, and is the pattern that native plants and animals have both adapted to and thrived until now.

The 2022 Cedar Creek Burn & The 1897 Organic Act

By Bob Zybach, PhD.

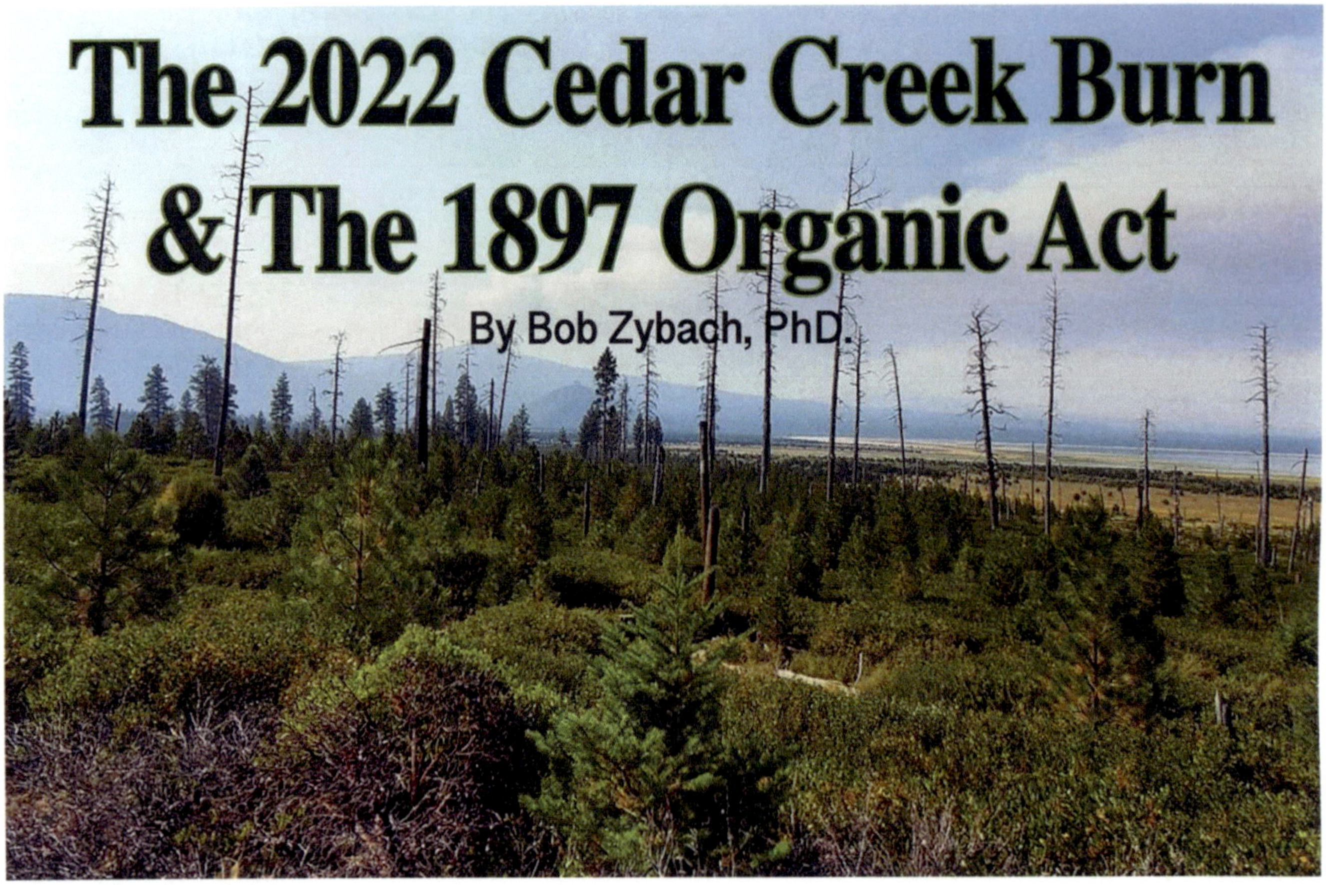

Fig. 1. Cedar Creek Fire Smoke Plume, September 9, 2022. Plume is from the west side of Maiden Peak, viewed from the south end of Davis Lake about 4:00 PM. Foreground snags are mostly ponderosa pine from the 2003 Davis Fire, which burned 23,000 acres. Some salvage logging had taken place but was stopped by litigation. Reproduction is mostly ponderosa and lodgepole pine, with some Douglas fir, which may have been planted among the natural seeding. Plume is blowing toward Bend. Photo by Rob DeHarpport.

The 2022 Cedar Creek Fire in Lane County and the Willamette National Forest in western Oregon, burned 127,311 acres. This number included 11,709 acres of so-called "critical habitat" (which the Forest Service calls "CH") for spotted owls ("NSO"), as defined by college professors (PhDs), litigation (EAJA), and the Endangered Species Act (ESA).

Local towns of Oakridge and Westfir were severely threatened by the wildfire, were blanketed with unhealthy smoke for nearly six weeks, and forced to evacuate entirely for four days while power was shut down.

This didn't need to happen. The wildfire was said to have been poorly managed by the Forest Service ("USFS") for an entire month before predictably exploding on an east wind in early September. By the time it was fully extinguished on November 22, taxpayers had spent more than 132 million dollars to employ more than 2500 firefighters that had operated dozens of helicopters and heavy equipment for the previous three months..

I visited the Cedar Creek Burn a few months ago with three friends, to view the results of two years of Forest Service projects following the fire. We had to get a special permit to pass a locked gate that was installed to keep the public away from the site. That seemed to be it: approximately two billion-plus feet of standing timber, rotting in place, primed to burn again -- only much hotter next time -- no apparent response, and the public is locked out.

The November, 2022 USFS BAER ("Burned Area Emergency Response") report doesn't mention the obviously increased risk of deadly wildfire, but states instead that: "Threats include additional loss of habitat in the fire area due to blowdown, mass soil movement, flooding, and insects and disease." But not wildfire?

Further: "A secondary issue includes determination if the proposed BAER stabilization treatments could affect spotted owl nest sites or result in disruption of nesting if conducted during the critical breeding season from March 1-July 15."

Say what? More than 125,00 acres of trees are dead, nearly 12,000 acres of so-called "CH" has been destroyed, and yet the government can't do anything about it for half the year because of SWO "critical breeding season?" How can this be written, reviewed, and publicly distributed, much less taken seriously?

Speaking in metrics and acronyms and producing voluminous busy-work reports seems to be the effective strategy. Also, the apparent SWO CH "critical breeding period" has apparently now been extended to the entire year, as there was no evidence at all of either USFS or spotted owl occupation at that time -- just a locked gate.

Cedar Creek Fire Timeline

The Cedar Creek Fire started by lightning strikes on August 1, 2022, on the Willamette National Forest, about 15 miles east of the towns of Oakridge and Westfir. According to the BAER report, it couldn't be extinguished as it "grew rapidly" in "inaccessible terrain," and then was "held at control features for around one month" until the east winds arrived -- as generally expected -- in early September.

It is currently unknown why smokejumpers weren't sent to extinguish the fire, but a helicopter with two "rappelers" flew over at 4:45 PM and reported it was about two acres in size and

Fig. 2. "Nature's Clearcut." View of increasing risk of wildfire the longer these snags remain in place. What had been large, growing plants filled with water were transformed in minutes by fire into massive amounts of pitchy, air-dried firewood. Unless something is done with these dead trees, the next fire through here -- as with all predictable Douglas fir reburns -- will be much hotter and cause more widespread damage than this fire. When snags fall before they reburn, they can also cause far more damage to the soil. Mark Cosby and Rob DeHarpport provide scale. Photo by author, October 28, 2024.

"burning in heavy timber midway up a 40% slope on a significant cliff with limited options for egress."

At that point the rappelers turned down the assignment, reporting the terrain was "too hazardous for safe access." Instead, a plan was developed to hike firefighters into the fire "during the next shift," while helicopters performed bucket drops to limit its growth. The fire was estimated to be about five acres on August 2 and grew to an estimated 500 acres on August 3.

By August 15, the fire had grown to 4,422 acres with 0% containment. All recreational trailheads and dispersed camping were closed both west and north of Waldo Lake. The following day, August 16, the Forest Service posted this statement, along with a three-minute video on its new Facebook public information page:

"Since it started on August 1st, the Cedar Creek Fire has been slowly getting bigger, moving east into wilderness and roadless areas, and south towards the bottom of Cedar and Black Creeks. Hundreds of firefighters have been deployed around the fire, creating primary and backup firelines. This effort is guided by a strategic planning process that combines the latest fire science with on-the ground observations and years of firefighting experience. To learn more about the strategic planning process and why it is being used at Cedar Creek, watch this short video."

On August 27, the fire had grown to 7,632 acres with 0% containment. The "strategic planning process" involved possibly the very first use of USFS "PODs":

"The District identified a contingency line based on a fuels assessment known as PODs (Potential Operational Delineations) well west of the incident. This line would provide a reliable fuel break in the event of a major east wind as well as any future fires that may threaten the communities of Oakridge and Westfir. While fire modeling and seasonality during the early stages of the incident did not indicate a strong likelihood that this line would be utilized, all the teams assigned continued to work on this contingency option."

On September 4 the fire began burning through the 1996 10,400-acre Charlton Burn in the Wilderness along the north side of Waldo Lake. The fire had never been salvaged or treated and was covered with fallen snags that burned very hot and sterilized the soil in much of the area.

By September 8, the fire had grown from 27,000 acres and reached over 73,000 acres according to news sources - however, the government reported an increase to only 33,000 acres, and growing to 52,000 acres on September 10.

On September 9, the USFS official online information source for wildfires, "InciWeb," reported the "Current Situation" as: "East winds, low humidities, and high temperatures will cause the fire to be active today. The highest activity will occur where lichen is present in trees and there is a high concentration of down wood. Where winds align with slopes, tree canopy fire and fire spotting are anticipated. E winds 15-20; gusts to 50."

Given this report, the Lane County Sheriff's Department made the decision to move Oakridge and Westfir into a "Level 3" evacuation, where residents are asked to leave by a given time and access roads are typically blocked. Powerlines were "de-energized" throughout the wind event and the duration of the evacuations, which were ordered for the following day.

Smoke from the fire moved into southwest Washington on September 10, and Seattle recorded the worst air quality of any major city in the world, with Portland coming in third out of the top 90, and Lahore, Pakistan sandwiched between the two. Seattle's "AQI" was 170 and Portland's was 152 (more on this later).

Sometime at this point the fire began burning into the infamous 1991 Warner Creek Fire scar, which had made national news for several years due to threatened legal actions and active anti-logging protests by "concerned environmentalists."

The 8,973-acre fire had cost $10 million to control and killed an estimated 180 million board feet (mmbf) of timber. Of this amount, only 35 mmbf were scheduled for sale, but two years of protests reduced that number to nine mmbf. Five years of protests and litigation later, no salvage had been completed, and the proposed sales number became only two mmbf. Then, 540,000 board feet were finally sold, and then the sale canceled. In September 2022, it all burned up.

Fig 3. Blair Lake. This was a popular recreation spot for fishing and boating before the Cedar Creek Fire. It is now gated off from the public and surrounded by dead trees. In earlier times the lake would have formed a natural firebreak, and it is unlikely that the large amount of second-growth trees that crowned would have been present in such large numbers. A few trees that survived the fire might be good as a seed source for future reforestation efforts. No information on reopening. Photo by author, October 28, 2024.

By September 11th, the fire had grown to 86,000 acres, more than 2,000 homes had been evacuated, and all highway access to the towns was blocked. Two days later the power was restored, and residents were allowed to return to their homes.

On September 12th, the wind had shifted to the northwest and smoke began affecting air quality in Sisters and Bend, while clearing the air in Oakridge for a few days.

On September 16th, the Central Oregon Daily News reported the fire had grown to 92,596 acres and videotaped Joan Kluwe, the "Public Information Officer for the Alaska Incident Management Team," saying: "We have 92 engines, 39 crews, 113 types of heavy equipment, and 19 helicopters working on the fire."

The total cost to "fight the fire, maintain and operate the machinery and take care of the personnel" to that date was given as $57,946,000.

On September 25th, the fire had grown to 114,104 acres with 20% containment.

By October 27th, the fire was 127,283 acres with 60% containment. Due to seasonal dropping temperatures and rising humidity, fire progression had nearly stopped, and the workforce had been reduced to two crews, one helicopter, two masticators, and four engines.

The following day, Fall rains and snow began, and finally, on November 22, the Cedar Creek Fire was declared "out."

1897 Organic Act

In 1897 Congress passed the "Organic Act" to manage and protect US public forestlands. The bill was signed into law by President William McKinley and has never been repealed.

The guiding principal of the Act remains fairly well-known to this time, and has been the theoretical basis to all subsequent USFS planning:

"No public forest reservation shall be established, except to improve and protect the forest within the reservation, or for the purpose of securing favorable conditions of water flows, and to furnish a continuous supply of timber for the use and necessities of citizens of the United States."

A lesser-recognized portion of the Act also states it was "for the purpose of preserving the living and growing timber and promoting the younger growth on forest reservations" -- and, in that regard, authority was given to "designate and appraise so much of the dead, matured, or large growth of trees found upon such forest reservations" for sale at "not less" than the appraised value, under the condition it couldn't be "exported" to another State or Territory.

In a nutshell, "living and growing timber" was intended to be "preserved," "younger growth" was to be "promoted," and a "continuous supply" of "dead, matured, or large" trees were to be sold at market value. "For the use and necessities of citizens of the United States."

In 1905, President Theodore Roosevelt, working closely with Yale-educated forester and close associate, Gifford Pinchot, transferred management of the Forest Reserves from the Department of the Interior to the Department of Agriculture, creating the US Forest Service.

Pinchot became the first Chief of the new agency and wrote that its mission would be: "Where conflicting interests must be reconciled, the question shall always be answered from the standpoint of the greatest good of the greatest number in the long run."

Theoretically, that has remained the mission of the Forest Service to the present time, and all subsequent forest planning laws and regulations are based on this vision and the 1897 Organic Act; including the "expanded purposes" of the 1960 Multiple-Use Sustained-Yield Act (MUSY) and the 1976 National Forest Management Act (NFMA).

Fig. 4. Spring Prairie. There are numerous oak trees, huckleberry patches, and beargrass meadows in the Willamette National Forest in the vicinity of Oakridge. These are strong indications the land was regularly occupied and burned by Molalla and Klamath people during early historical time -- and likely for many generations before. Several of the surviving trees around the prairie's perimeter might be good seed sources for future reforestation efforts. Photo by author, October 28, 2024.

The 1960 Act included the considerations of range, water, recreation, and wildlife, along with timber, as renewable resources, and the 1976 Act changed forest planning by requiring the Forest Service to use an "interdisciplinary approach" to resource management -- and to include "public involvement in preparing and revising forest plans."

Local Effects: Smoke & Jobs

Bryan Cutchen is the Mayor of Oakridge, and he says wildfire smoke for the past four years, from the Cedar Creek Fire and others, has done great damage to the town's recreation and hospitality businesses. Others in the community point to the 2015 and 2017 fire years as a major beginning to these events.

The combination of reputation and reality, smoky air, road closures, and burned campgrounds have kept people away from Oakridge in recent years. The discomfort and poor health associated with breathing smoke and the reduced visibility and public access following the fires has caused serious economic damage to the community.

Rob DeHarpport was the third generation of his family to live in Oakridge. His grandfather worked for the Forest Service and his father owned a store in town, and that's where he graduated from High School in 1977 and has lived most of his life. After graduating, he spent a summer fighting fires with the Oakridge Ranger District, and from 2012 to 2015 he became Mayor of Westfir, before retiring and moving east of the Cascades.

During the years he was growing up in town and for many years after, he can only recall a few times when Oakridge became smoky -- typically for a day or two when smoke would drift into town from field burning in the Willamette Valley or from fall slash pile burning in the woods.

In 1975, when Tim Bailey began his career as a Forester for the Oakridge District Ranger Station, the town had eight gas stations, eight bars, two jewelry stores, two florist shops, a movie theatre, bowling alley, skating rink, and 400 people working at the Pope & Talbot Mill. Today there are three gas stations, up from just one a few years ago and a one-man firewood cutting business.

Bailey's memories are similar to DeHarpport's regarding the general lack of smokey summers and falls. At that time Oakridge considered itself the "Tree Planting Capital of the World" and still has an annual festival to recall those years.

When the 9,000-acre Shady Beach Fire burned in 1988, the 195 mmbf of snags and surviving trees were quickly sold and harvested -- creating an estimated 200 local jobs, and the land was then planted. Today, it is a healthy stand of 35-year-old trees nearing merchantable size.

Fig. 5. Industrial Plantation. Following WW II, most reforestation in the Douglas Fir Region has been to industrial standards with a focus on fiber production and future jobs. Planting seedlings in rows following logging or wildfire, or converting old meadows, prairies, and berry fields to trees results in a significant loss of biodiversity and wildlife habitat, while creating increased opportunity for deadly crown fires. Russ Sapp and Dave Sullivan in Preacher Creek plantation in Lincoln County. Photo by author, April 5, 2024.

Bailey wasn't directly involved in the Shady Beach Fire but was the silviculturist for the subsequent Warner Creek Fire. The starkly different histories of the two fires characterized the

Fig. 6. Crevice Creek Plantation. This 60-year-old industrial-type plantation along Crevice Creek experienced 100% mortality. Despite the relative small size of these snags, they will remain very flammable and an increasing risk of wildfire if they are not removed. Although they no longer have value for lumber production, they can likely be used to produce commercial fuel pellets for several more years. Note the two-year understory of invasive grasses and brush that have replaced the green trees and shaded ground. They will form volatile flash and potential ladder fuels for the thousands of tons of dry, pitchy firewood that has been created. Photo by author, October 28, 2024.

Date 2022	Oakridge	PM2.5	Eugene	PM2.5	Bend	PM2.5	Sisters	PM2.5
Sept. 1	110	39.1					**8,817 Acres**	
Sept. 3					110	39	**16,303 Acres**	
Sept. 6	141	51.9					**18,143 Acres**	
Sept. 9	**211**	**160.6**					**33,099 Acres**	
Sept. 10	**341**	**290.6**					**51,814 Acres**	
Sept. 11	194	138.8			124	45	154	60.5
Sept. 12					**213**	**162.7**	**172**	**96.4**
Sept. 13					154	62.2	124	45
Sept. 14					147	54	**92,595 Acres**	
Sept. 17	173	98.5					**110,361 Acres**	
Sept. 18	171	93.5						
Sept. 19	**283**	**233.4**	**144**	**53**				
Sept. 20	174	99.4						
Sept. 21	**227**	**176.8**					**113,637 Acres**	
Sept. 23	173	99						
Sept. 24	156	66.1						
Sept. 25	155	63.9						
Sept. 26	153	58.9						
Sept. 27	160	72.2						
Sept. 30	**227**	**176.3**					**120,546 Acres**	
Oct. 1	152	57.7						
Oct. 2	138	50.7						
Oct. 3	158	68.2						
Oct. 4	168	87.8						
Oct. 5	154	61.1						
Oct. 6	**213**	**163.1**					**120,983 Acres**	
Oct. 7	**216**	**165.5**						
Oct. 8	**254**	**204**						
Oct. 9	**228**	**178.1**						
Oct. 10	**285**	**235.4**	**161**	**75**			**122,210 Acres**	
Oct. 11	182	116.2						
Oct. 12	**314**	**263.6**						
Oct. 13	**297**	**246.9**	**115**	**41.1**				
Oct. 14	**362**	**312.3**	**100**	**35.2**			**123,498 Acres**	
Oct. 15	**333**	**282.9**	**132**	**48.1**				
Oct. 16	158	69.3						
Oct. 17	**206**	**155.3**						
Oct. 18	**359**	**309**	**102**	**35.8**				
Oct. 19	**380**	**330.4**	**156**				**126,222 Acrea**	
Oct. 20	**303**	**252.4**						
Oct. 21	164	81.3					**127,283 Acres**	

Table 1. Cedar Creek Fire Air Quality Index (AQI) and PM2.5 Particulates

dramatic change in USFS management philosophies that took place with the arrival of spotted owls in the Douglas Fir Region in 1990.

The Table with this article shows all Air Quality Index (AQI) numbers and PM25 measures for Oakridge, Eugene, Sisters, and Bend during the smokiest days for those communities during the Cedar Creek Fire. All days that reached an AQI of 100 are listed; those above 200 are bolded; and all above 300 are also made red. The highest PM25 readings for those days are also given, and because Sisters was only impacted for a few days, progressive fire sizes are also listed in that column.

These data were provided by Ryan Porter, who authored the Oregon Department of Environmental Quality (DEQ) report on Oregon statewide air quality for 2022. DEQ AQI numbers are based on a scale of 0-50 (Good); 51-100 (Moderate); 101-150 (Unhealthy for sensitive groups); 151-200 (Unhealthy); 201-300 (Very Unhealthy); and 301-500 (Hazardous).

Compare these numbers to those given earlier for September 10 in Seattle and Portland to better understand the localized severity of this problem.

PM25 numbers represent particulate matters in the air that are 2.5 micrometers or less in diameter. Readings above 35 are considered "unhealthy for short-term exposure," and the National Institute of Health states that "people with respiratory diseases" account for a "large proportion" of "non-accidental deaths" in the US. The stated reason is because PM25 pollution "causes asthma, respiratory inflammation, jeopardizes lung functions, and even promotes cancers."

A recent public health study concluded that more than 10,000 people a year in the US die from wildfire smoke pollution, and recent University of California research indicated PM25 exposure from wildfire smoke likely killed 55,000 Californians alone in the 11 years from 2008 to 2018.

Recommendations

What can be done to fix this deadly, costly mess? It took 35 years for the Willamette National Forest to transform itself from an increasingly unpopular actively-managed clearcutting marketplace to a seasonal inferno, causing massive air pollution for weeks, killing millions of wildlife and old-growth trees, bankrupting local businesses, ruining the scenery, ending recreation, blocking our public roads with gates, and charging US taxpayers billions of dollars to make it all possible.

Even though significant fault can be found with the design of previous timber sales and reforestation projects, a course correction rather than a complete reversal would have been a more reasonable approach to spotted owl politics. Subsequent planning processes and environmental lawsuits have been allowed, and even encouraged through favorable rulings and legislation, to transform our once beautiful and productive National Forests into ugly, costly, and deadly firebombs. How to fix?

These photos, and millions of others, clearly show the problem. Passively managed conifer forestlands and plantations don't somehow magically transform themselves into "critical habitat" for obscure animals, or into 400-year-old stands of trees teeming with "biodiversity." No, some grow old, they die, and if nothing is done, they burn and are replaced with a new generation of plants, often the same species, but in different configurations.

Fig. 7. Second-growth Douglas fir progeny competing with ancestral old-growth for sunlight, water, and soil. The original wide-spacing of the older trees is indicated by the large stobs that remain from lower limbs lost to competitive shading along their trunks. These remaining old-growth are now at increasing risk of a crown fire, in addition to direct competition from younger trees. Removing the second-growth to protect the old-growth could be accomplished at a good profit, as well as greatly reducing risk of crown fire and resulting loss of remaining old-growth. Photo by author, April 5, 2024.

The real problem began in the 1970s and the deep ecology movement, when the woods became "wildlands" and people were pathogens and the enemies of "nature." Then the academic myth of the "healthy forest" containing "big, dead, standing trees," large logs scattered around on the ground, and a "multi-layered canopy of diverse species" was invented.

To these "New Forestry" ecologists, life appeared to exist in a terrarium, where the ideal was a "non-declining, even-flow, naturally functioning ecosystem" teeming with endangered species and incredible biodiversity. Their computers "proved" this ideal could be achieved "again" -- if only we'd eliminate people, their matches, and their wildlife-disturbing road systems.

Environmental activists, their lawyers, and politicians attempted to transform this fiction into a vision for our public lands, and it has never worked. Instead, death, fire, smoke, and poverty has resulted -- as predicted by actual foresters and knowledgeable residents.

If this fire had happened 40 years ago, in 1982 instead of 2022, many things would have been different. Tens of thousands of fewer acres would have burned; millions more wildlife would have lived; thousands more old-growth would have survived; there would have been only a fraction of the smoke for far fewer days; and the wood would have been immediately salvaged -- and at a profit to local families and to American taxpayers. Then the land would have been reforested as well as possible, based on the best intentions at that time.

A concerted effort to return to the common sense purposes of the 1897 Organic Act, Pinchot's fundamental mission statement, the 10:00 AM Policy, and the 1962 MUSY would be wonderful. That would be my best recommendation.

This systematic destruction of our public forests and rural communities needs to stop -- and should have been stopped long ago.

A sawmill for plantation and old-growth thinnings and a pellet mill for processing fire-killed trees could be located at the old Pope & Talbot Mill site that Oakridge currently owns. All that would be needed would be a long-term USFS contract, such as established the town of Westfir or the Sustained Yield Units that followed WW II. Even the CCCs from the 1930s could be a model for earning citizenship if modern-day US citizens really won't do "this type of work."

There is much work to be done, it will take thousands of people many years to do it, and it would be a lot cheaper and healthier and far safer and more beautiful, profitable, and productive than what we have now. Or we could keep doing this, and continue to "let nature take her course" and let the lawyers do the explaining.

FOREST REFORESTATION & RESTORATION: AN ANTHOLOGY

FOREST REFORESTATION & RESTORATION: AN ANTHOLOGY

FOREST REFORESTATION & RESTORATION: AN ANTHOLOGY

Made in United States
Troutdale, OR
03/25/2026

47698612R00088